THE COMPLETE GUIDE FOR GARDEN, RANCH, AND FARM

PALMETTO
PUBLISHING
Charleston, SC
www.PalmettoPublishing.com

Copyright © 2024 by Rodney W. Bovey

All rights reserved

The author has made every effort to ensure the accuracy of the information within this book was correct at time of publication. The author does not assume and hereby disclaims any liability to any party for any loss, damage or disruption caused by errors or omissions, whether such errors or omissions result from accident, negligence, or any other cause.

No part of this book may be reproduced or transmitted in any form and by any means, electronic or mechanical, including photocopying, recording or by any information storage and retrieval system, without written permission from the author.

Hardcover ISBN: 979-8-8229-4642-2
Paperback ISBN: 979-8-8229-4643-9

THE COMPLETE GUIDE FOR GARDEN, RANCH, AND FARM

The History, Science, and Art of Growing Food

RODNEY W. BOVEY

PREFACE

The Complete Book of Farming and Gardening is written for the layman and professional alike. The book is written with limited technical language about agriculture procedures and production giving the reader a clear and interesting picture of modern agriculture. Most people purchase their food at the local grocery store and may have no idea for the need of basic agriculture science and funding and the great effort scientists, educators, ranchers and farmers make in providing our food supply. This book attempts to cover all the important aspects of agriculture including early to modern history, the importance of soils, water, weather, native and domestic crops and livestock, early and modern technology, and food science. It also includes pest and pollution management, farm machinery, marketing policies, training and much more. Concluding remarks and future needs are given to provide food for a growing projected population of 9 billion people by 2050.

The author wishes to dedicate this book to my father, William "Bill" August Bovey, who was a good farmer, businessman, community leader and father. He was my mentor growing up and provided well for his family and set a high standard for me, my sister Jacque, himself and others.

ACKNOWLEDGMENTS

The author wishes to extend a special thanks to A. Jenét Ardoin (Beck), Office Science Associate, Ecosystem Science and Management, College of Agriculture and Life Sciences, Texas A&M University, College Station, Texas for editing and typing the original manuscripts.

A special thanks to Brooke Livingston for editing and typing the final copy of this book and to Austin Bovey for designing the book cover, contacting the publisher and finishing special assignments of the project.

The author thanks Texas A&M University, Department of Ecosystem Science and Management, for providing funds, office space, secretarial help to write and complete this publication and the Dripping Springs Texas Community Library for providing space and encouragement to finish the book.

A special thank you to Dr. Scott Senseman, Director, Oklahoma State Agricultural Experiment Station, Oklahoma State University, Stillwater, Oklahoma, for reviewing the entire manuscript.

CONTENTS

EARLY TO MODERN AGRICULTURE

The purpose of this manuscript is to show the difficulty and huge effort in plant and animal improvement and supplying food, clothing, and shelter for a large growing population. The task is never complete because new pests or environmental conditions constantly emerge that require new or improved genetics, pest control, fertilizer application, and improved nutrition, animal care, and production. The search for improved and adapted plant and animal sources has continued since early man began domesticating these essential agricultural products.

Recent technological advances have helped our cause, but we must be ever cognizant to maintain and preserve the genetic integrity of our wild and domestic plants and animals for future use.

The problems in providing information and technology to maintain agriculture is not all negative because it provides many jobs for farmers, ranchers, foresters, research scientists, industry, teachers, machinery dealers, agricultural engineers, herbicide and pesticide personnel, fertilizer experts, food and seed personnel, and many others.

Work in agriculture is very interesting and requires expertise in many fields. It is rewarding to see a new crop variety developed, the registration of a new crop pesticide or improved tree harvesting technique.

Early History

Levack et al. (2007) indicated food production made civilization possible. For thousands of millennia humans did not produce food during the Paleolithic Age or Old Stone Age because people scavenged for wild food. By the end of the last Ice Age about 15,000 years ago as the Earth warmed, cereal grasses spread and hunter-gatherers learned to collect these wild grains for food (Levack et al. 2007). Domestication of plants and animals (pigs, sheep, goats and cattle) soon occurred. Levack et al. (2007) indicated first signs of domestication occurred about 8900 B.C.E. in southwest Asia with domestication of pigs around 7000 B.C.E. and domestication of cattle, goats, and sheep around 6500 B.C.E. Planting and harvesting of cereal grains occurred sometime between 8000 and 7700 B.C.E. (Levack et al. 2007).

Thompson (2010) indicated different people in different places domesticated different plants depending on what they found around them and the growing conditions they faced. Nomadic tribes people observed grasses and other useful plants spreading into patches after fires had destroyed other herbaceous and woody plants. Domestication and cultivating plants developed slowly over the millennia, as well as domestic livestock. Retreat of the glaciers marks the beginning of the Mesolithic or Middle Stone Age. The Neolithic or New Stone Age occurred next as humans began to cultivate land. As the farming economy developed starting about 8000 years ago, sheep and goats became domesticated that controlled shrubs and utilized grasslands. Weeds (unwanted, destructive plants) and wild birds feeding on cereal grains became problems for growing crops (Thompson 2010). Weeds, insects, plant diseases, nematodes, rodents, and other pests and predators of crops are still big management problems in modern agriculture.

Modern agriculture is composed of developing varieties of crops that are genetically superior to older varieties by increased resistance to drought and pests and improved yield and human nutrition. Genetically-modified crops may be produced to tolerate certain herbicides and pesticides, improve yield, nutritional qualities, and environmental constraints. Similar trends have occurred with domestic livestock to improve meat production, disease resistance, heat or cold resistance, egg, pelt, or milk quantity or quality. Thompson (2010) reported that wheat, barley, maize, rice, rye, millet, and sorghum (the staple grain crops of the world), soybeans, peas, lentils, beans, sweet potatoes, potatoes, squashes, dates, and bananas all originated from our Neolithic ancestors. Cultivated fruits, vegetables, cereals, and many other crops originated very early before science played an active role.

Thompson (2010) stated that the Soviet geneticist Nikolai Vavilov (1887-1943) was a pioneer in documenting the importance of genetic variation in the conservation of crops and inspired the foundation of numerous seed banks in the late 20th century. Vavilov and his collaborators identified eight centers of crop origin or crop diversity. They are China, Southeast Asia and the Pacific Islands, Central Asia, Near East, Mediterranean, Ethiopia, Mesoamerica, and South America and the Andes.

China was one of the world's most significant centers of crop domestication in antiquity. India was important in introducing tropical crops and vegetables, including rice, lentils, cucumbers, aubergines, mangos, and other tropical fruits. Villages in Mexico cultivated maize, tomatoes, upland cotton, beans, squash, chilies, sweet peppers, and numerous other crops and vegetables. Thompson (2010) indicated that the entire gene pool of cereals, vegetables, and other plants on which humans depend was generated in a few limited regions of the world. Genetic diversity from which crop improvements will be made in the future is restricted to the plants growing in the same regions.

Some crops were not cultivated in regions of similar climate, mainly because many hunter-gatherers remained hunter-gatherers. Most skilled cultivators could not achieve success until seeds produced seedlings and some crops were just not amenable to cultivation.

Northern Europeans in the Neolithic period initially rejected farming as it spread throughout the continent. This was determined by the adoption or rejection of ornamentals - certain types of beads or bracelets worn by different populations (d'Frrico and Vanhaeren 2015). First farmers came to Europe 8000 years ago and the spread of ornaments were linked to farmers - human-shaped beads and bracelets composed of perforated shells - stretching from eastern Greece and the Black Sea shore to France and from the Mediterranean Sea northward to Spain. Hunters and foragers in the Baltic area resisted the adoption of ornaments worn by farmers during this period.

However, Britons may have discovered a taste for bread thousands of years earlier than previously thought (Allaby 2015). The practice of planting and harvesting cereals occurred about 12,000 years ago in the region where Europe meets Asia and slowly spread across Europe. Britons didn't adopt agriculture until 6000 years ago, but by analyzing sediment samples from the Bouldnor Cliff underwater site off the Wight Isle it was found that wheat was present 8000 years ago. This finding suggests trading between hunter-gatherers (Britons) and farmers (Mesolithic people).

Ebeling (1979) indicated today hunting and gathering are activities modern man, practically Cro-Magnon genetically, now widely considers to be recreation in developed countries. During the two million years to the dawn of man's cultural evolution, population growth was very slow. This was attributed to the exigencies of the nomadic lifestyle of the hunter-gatherers and not because of an inadequate food supply.

However, about 10,000 years ago human population growth occurred and adoption of farming practices (technological change) was necessary to sustain growth. By 8000 B.C. human population had expanded from tropical to temperate and arctic latitudes in both hemispheres (Ebeling 1979). Big game animals had become scarce and an increased quantity of small animals and vegetables were being consumed. Fire also became important to clear land for growth of animals, cereal, and other crops.

The Neolithic farmer could provide food for his own family and several others (Ebeling 1997). The hunter-gatherer lifestyle may have been preferred to the backbreaking toil of farming. However, mechanized equipment and fossil fuel eventually provided some relief in developed countries. Many would argue, although necessary, that modern mechanization and farming methods have resulted in enormous environmental degradation and loss of natural resources.

Dubos (1980) pointed out that ever since the development of agriculture in the Neolithic Period, the immense majority of humans have lived in environments they have transformed. As a result, few humans really desire to inhabit the wilderness permanently and even fewer could survive very long in it. The word wilderness occurs about three hundred times in the Bible and all its meanings are derogatory (Dubos 1980). Man has struggled against environments to which it could not adapt and has shunned the wilderness or has destroyed much of it the world over. In the New World, Europe and the Orient, humans feared the wilderness; but it changed to admiration probably as soon as dependable roads gave confidence to safe and comfortable quarters (Dubos 1980). Although many humans may not live in the wilderness, it is extremely important to trap energy, hold genetic diversity of plants and animals, provide protection, recreation, beauty, resources and other benefits.

Dubos (1980) further states that human intervention into nature was many times destructive even before agriculture and herding. Fires were set by pre-agricultural people to kill trees and plant crops. Ecological disasters of inhabitants of Mesopotamia, South America, China and other areas by deforestation and overcropping have caused damaging soil loss and wildlife loss or extinction. It is human nature, however, to modify the environment to suit human survival and needs. Man was given dominion over the earth and its plants and animals. Agriculture will need to support 9 billion people by 2050. Our task will be to develop the required technology by improving or doing little harm to the environment.

Gregor Johann Mendel (1822-84), the father of plant genetics, made controlled crosses between peas with different characteristics. Mendel deduced the basic laws of inheritance by counting the numbers of the different types of offspring that were produced (Thompson 2010). Thompson (2010) indicated that in 1830 the Academy of Science at Haarlem University offered a prize to anyone that could prove sexuality in plants. Seven years later the prize was awarded to the German botanist Karl Friedich von Gärtner, who proved sexuality in plants.

Mendel showed that characteristics in plants are inherited from generation to generation according to fixed laws (Thompson 2010). Mendel opened the way to more informed and logical approaches on breeding new and improved varieties of plants. In the meantime farmers, nurserymen, seedmen and others became plant breeders following Karl Friedich von Gärtners declaration of sexuality in plants. Thompson (2010) asks why should sex be so important in the making of a seed? The answer is that it creates offspring composed of mixtures of their parents' genes in a multitude of different combinations and this diversity is the key to survival when disaster strikes or new opportunities arise. The people and the genetic mechanisms responsible for our vast array of crops and varieties will be discussed in detail later.

Gardner (2001) published a 1088 page book on *Traditional American Farming Techniques* that was originally published in 1916. The accumulation of knowledge at that time was overwhelming and attempted to be a ready reference on all phases of agriculture for farmers of the United States and Canada. It discussed soils and soil management, farm crops, horticulture, forestry and floriculture, dairy, animal husbandry, farm buildings and equipment, farm management, plant and animal diseases, insects and their control, home economics and agricultural education and more.

The *Yearbooks of Agriculture* were published annually from 1886 to 1992 by the United States Department of Agriculture and were a great source of information. Each book usually contained specific subjects like grass, trees, crops, insects, plant diseases, marketing, water, etc. In addition, many societies were established in the last 50 to 100 years that publish popular and scientific articles and books on weed and pest management, new horticulture, agronomic and pasture crops, animal science, use and care of machinery and many other subjects as a result of scientific research, extension demonstrations and individual contributions. Markle et al. (1998) lists 691 field and forage crops grown in the United States. A description of each crop is given, the number of acres grown, its use, where it is grown, and classification by EPA. Much data is also available online.

The Restless Earth

In review of several books on earth science and personal experiences, it is clear that the earth is not a static entity but an ever dynamic planet subject to change (Calder 1972; Nebel 1987; Spencer 2003; Miller 2005). Many humans view the earth as a stable and unchanging world. However, earthquakes, volcanoes, mountain-building and movement of huge plates of our planet's outer shell are vivid reminders of change. Agricultural problems are also augmented by storms, drought, flooding, loss of topsoil, pests, predators and disease. Unfavorable weather is a constant threat to farmers and ranchers when dealing with crop and animal production.

Since a very small percentage of our people in the United States and Canada are farmers and ranchers, most people in urban areas feel far removed from food production and the ecological difficulties. However, there are many misinformed critics willing to correct the experts on pollution, climate change, ecological correctness and land use. It is positive that the population is concerned with climate change, agricultural ecology and environmental problems to keep the U.S.A. safe, healthy and prosperous. However, too much debate and accusations by the uninformed is unnecessary and destructive. Many ecological and environmental problems are exaggerated and overstated causing unwarranted alarm, cost, bad feelings among opposing parties and overregulation.

Not many years ago humans seemed unaware and unconcerned about our environment. Being aware is great and support for a clean, healthy ecosystem is welcomed. World-renowned expert Sir John Houghton (1997) explored the scientific basis of global warming and likely impacts of climate change on human society. Houghton (1997) indicated the important greenhouse gases which are directly influenced by human activities are carbon dioxide (CO_2), methane (CH_4), nitrous oxide (N_2O), the chlorofluorocarbons (CFCs) and ozone (O_3). The most important greenhouse gas is water vapor, but change in the atmosphere is not caused directly by human activities. With the advent of increased human activity, CO_2 concentrations are increasing at high rates from about 280 ppmv around 1700 to 360 ppmv by the year 2000. Global concentration today is about 410 ppm - a small increase. Methane in the atmosphere is much less than CO_2, <2 ppmv. It occurs mainly from wetlands. Methane adds to the greenhouse effect. It is destroyed in about 12 years (much shorter lifetime than CO_2) through chemical reactions in the atmosphere. Nitrous oxide occurs at about 0.25% in the atmosphere and is another minor greenhouse gas. The CFCs are man-made chemicals and remain in the atmosphere a long time. They destroy ozone. Ozone is formed through the action of ultraviolet radiation from the sun on oxygen molecules. CFCs reduce ozone if in the atmosphere. One chlorine atom can destroy many molecules of ozone. Houghton (1997) also indicated population growth is a problem likely to increase and cause disparity in wealth between developed countries and undeveloped countries, deplete many resources and cause global insecurity. Houghton's evaluation is magnificent and comprehensive; but, as stated, future predictions are only predictions.

Miller (2005) reports that there is considerable evidence that the earth's troposphere is warming partly because of human activities. Impacts of CO_2 and other greenhouse gases from human activities may raise the average global temperature of the atmosphere near the earth's surface. Spencer (2003) supports a similar view to Houghton (1997) and Miller (2005). There is still ongoing debate over the influence of greenhouse gases on global warming.

Glover and Economides (2010), however, state that 90% of our energy is a hydrocarbon energy mix (38% oil, 23% gas and 24% coal) with nuclear and hydroelectric power practically completing the remainder. Glover and Economides (2010) further state that alternative energy projects such as wind power, alcohol production from biomass and solar power are highly expensive, unpredictable and contribute little to satisfy energy needs. Holechek and Sawalhah (2014) argue that oil is a superior energy source; but when fossil fuels are depleted aggressively, developing alternative sources and implementing energy conservation will be required. By 2035 forecasts still suggest an overwhelming dominance by use of fossil fuels (Glover and Economides 2010).

Glover and Economides (2010) further suggest Europe, America and Western civilization, especially in the last century, have been able to develop economic prosperity and thrive; but the politics of energy has become a tool for social engineering and claims of big government to overstate the effects of global warming and climate change. Increasing big government by centralizing social policies eventually erodes personal choice and individual freedom. Getting oil or food from the field to the consumer is not easy and requires much energy and effort as we shall see in the pages that follow. It must be supported by government and the people.

Another commodity as valuable as oil is water. Much of our food supply depends upon irrigation and adequate rainfall.

Water Rights

Water is essential to grow crops, fish, land animals, to mine, to frack for oil and to sustain life. Glennon (2002) stated that in the 18th century the American colonists borrowed the English law of riparian water rights that allowed owners of land abutting lakes, rivers, and streams natural flow of water without deviation or alteration of the watercourse. The rule, however, eventually threatened to retard economic development as the country began to industrialize. This eventually created the prior appropriation doctrine which essentially states "first in time, first in right" and arose in the gold mining camps. Surprisingly this system endures today by allocating enormous quantities of water to senior appropriators to grow crops and livestock. Glennon (2002) further reports that such a system deprived others (Native Americans) of sometimes needed water. To alleviate this problem the federal government has constructed dams in suitable locations to control flow to farmers, municipal and industrial areas. This sometimes results in moving huge quantities of water long distances to satisfy water needs.

Energy Needs and Food Processing

Price (1980) indicated that the energy use in agricultural production relative to other sectors is less than 5% in the United States. However, consumption in the total food and fiber system is over 20%. Even though total consumption in agricultural production is low, it is very essential. Providing fuel on farms at peak times is critical. Agriculture is the only major sector utilizing solar energy on a broad scale. Through photosynthesis, plants convert the sun's energy to a ready storable and usable form of food, feed and shelter (lumber).

The products of agriculture provide energy needs for humans and nutrients and feed grains and legume crops provide energy and food needs for livestock.

Considerable additional energy, however, is required to transport, process, package, market and store food for the end user. Although this requires large quantities of energy, it has allowed the food processing industry to provide high quality packaged products for the consumer.

The United States has rich soil, favorable climate and the agricultural technology to provide food resources for our needs and for many foreign countries. Growing crops, especially for conversion to ethanol, is being done on a large scale. However, the net gain in energy for the cost involved is questionable. Also availability of oil from recent fracking has at least temporarily changed the dynamics of energy availability. Diverting corn and other crops from human use to ethanol production has increased food costs and availability, especially for developing countries.

Distinguished professor Dr. Norman Borlaug (deceased), Soil and Crops Science Department of Texas A&M University, College Station, Texas actually worked with scientists and graduate students in Central America, the Mid East and Africa to help improve their wheat varieties for greater production and self-sufficiency. Recent publication of "Yield Gains in Major U.S. Field Crops" is available (CSSA special publication 33, 2014).

Training for Agriculture

There is tremendous opportunity for careers in agriculture. These include consulting, research scientists, teachers, extension personnel, industry jobs, agribusiness, farmers, ranchers, food processors, agricultural engineers, and all kinds of workers and assistants.

Growing up on a farm is helpful but not essential. Being involved in 4-H, FFA (Future Farmers of America) and helping on the family farm are very useful and help direct young people to decide if agriculture is a desired career.

The list of opportunities for careers is almost endless and includes environmental science, biology, biochemistry, plant and animal husbandry, plant and animal physiology, ecology, agronomy, range science, horticulture, agricultural engineering, veterinary science, forestry, agricultural journalists and reporters, food science, hydrology and on and on.

Many farmers and ranchers train in business since many that have grown up on ranches or farms are equipped to handle the agriculture techniques and technology but lack business skills. Others without early exposure may want to be an apprentice under the guidance of an experienced farmer or get college training. Unless the apprentice has good financial support, getting started and staying in business may be difficult.

Avery (2002) indicated that the next 30-40 years will see the largest global need for food in human history. No one is better prepared to take advantage of the global farm market than North American farmers. We have the best land, the best research network, the best techniques, the best infrastructure and the best transportation system. American farmers and ranchers have a tough job physically, mentally, and financially. Farming and ranching have become targets for environmental activists, and the lack of free trade in agricultural products inflate prices and depress growth.

In spite of many problems, people who love the land and want to be independent will accept some sacrifice to be free and self-directed.

Chapter 1. Literature Cited

Allaby R (2015). Wheat Discovery Points in Ancient Trading Links. Retrieved from http:/www.laboratoryequipment.com/news/2015/o2wheat_discovery. Accessed 27 February 2015

Avery A (2002). *The Changing Role of U.S. Agriculture in World Food Production*. Texas Plant Protection Conference. College Station, Texas. 14 p

Calder N (1972). *The Restless Earth*. The Viking Press. New York. 152 p.

d'Frrico F, Vanhaeren M (2015). Neolithic Europeans Didn't Enhance Farming. Retrieved from http://www.laboratoryequipment.com/news/2015/04northern_neolithic. Accessed 9 April 2015.

Dubos R (1980). *The Wooing of Earth*. Charles Scribner's Sons. New York. 183 p.

Ebeling W (1979). *The Fruited Plain:The Story of American Agriculture* (pp. 1-18). University of California Press. Berkeley and Los Angeles, California.

Gardner FD (2001). *Traditional American Farming Techniques: A Ready Reference on All Phases of Agriculture for Farmers of the United States and Canada* (Originally published in 1916 by L.T. Meyers as *Successful Farming*). Lyons Press. Guilford, Connecticut. 1088 p.

Glennon R (2002). *Water Follies: Groundwater Pumping and the Fate of America's Fresh Waters*. Island Press. Washington, D.C. 314 p.

Glover P, Economides M (2010). *Energy and Climate Wars: How Naive Politicians, Green Ideologies and Media Elite are Undermining the Truth About Energy and Climate.* The Continuum International Publishing Group. New York, New York. 231 p

Holechek J, Sawalhah M (2014). *Energy and Rangelands: A Perspective.* View Point. Rangelands, 36:36-43.

Houghton J (1997). *Global Warming: The Complete Briefing* (2nd ed.). Cambridge University Press. Cambridge, United Kingdom. 251 p.

Levack B, Muir E, Veldman M, Maas M (2007). *The West: Encounters & Transformations* (2nd ed., pp. 13-14) Pearson Longman. New York, New York.

Markle G, Baron J, Schneider B (1998). *Food and Feed Crops of the United States: A Descriptive List Classified According to Potentials for Pesticide Residues* (2nd ed.). Meister Pub. Willoughby, Ohio. 517 p.

Miller GT (2005). *Sustaining the Earth* (7th ed.). Thompson Learning-Brooks/Cole. Pacific Grove, California. 115 p.

Nebel BJ (1987). *Environmental Science. The Way the Worlds Works.* Prentice-Hall. Englewood Cliffs, New Jersey. 671 p.

Price D (1980). Where Farm Energy Goes. In J. Hayes (ed.). *Cutting Energy Costs: The 1980 Yearbook of Agriculture.* U.S. Department of Agriculture Superintendent of Documents. U.S. Government Printing Office. Washington D.C.

Spencer E (2003). *Earth Science: Understanding Environmental Systems.* McGraw-Hill. New York. 518 p.

Smith S, Diers B, Specht J, Carver B (eds.) (2014). *Yield Gains in Major U.S. Field Crops.* CSSA Special Publications, 33. Crop Science Society of America, Inc. Madison, Wisconsin. 487 p.

Thompson P, Harris S (2010). *Seeds, Sex and Civilization: How the Hidden Life of Plants Has Shaped Our World.* (pp. 16-30, 33-68, 112-120) Thames & Hudson. London.

SOILS

Introduction

Wiley (1895) in the 1895 Yearbook of the United States Department of Agriculture indicated that soil was no longer regarded as dead and inert matter but was known to be permeated with a living mass. They recognized the importance of soil bacteria and their role in providing nitrogenous food for plants and supplemental fertilization on areas of long-term cropping. Other articles discussed alkali soils, humus and soil fertility and cultivation of soil as well as many other articles on crops, food production, plant diseases, weeds, forestry, and livestock.

Good soil resources and water are basic to any prosperous nation and basic to food production as well as good management of those resources. Good soil management includes conservation and maintenance of its integrity, structure, and function when producing animal and plant products. Recognition of good management has been a part of agriculture in the United States and many other nations for many years (Gardner 2001). Good management also includes care and repair of machinery, cost of production, size and diversity of the farm in relation to efficiency, labor and capital needs, marketing and record keeping. Good management also includes care of water resources. Pollution of soil and water must be avoided as well as prevention of soil loss by air or water movement.

Soil Description

Simonson (1957) describes soil as a continuous layer over the land surface of the earth, except for the steep and rugged mountain peaks and the lands of perpetual ice and snow. It varies in depth but links the rock core of the earth and the living things on its surface. Soils vary from deep to shallow and red to black in color but all consist of mineral and organic matter, water and air. Soil has a profile - a succession of vertical layers known as horizons. Horizons differ in one or more properties, such as color, texture, structure, consistency, porosity, and reaction.

Soil Profile

Most soil profiles include three master horizons identified as A, B, and C. Horizons A and B are usually the major profile but sometimes can be eroded away. A-C profiles are possible. All Master horizons may be subdivided into A_1, A_2, A_3, B_1, B_2, B_3 or other divisions. The subdivisions provide clues to soil formation and management. The A horizon is the surface layer and contains plant roots, bacteria, fungi and small animals.

The soil surface is abundant in life and contains the most organic matter. The A horizon is subject to loss of clay, iron, aluminum oxides, and nutrients if in high rainfall areas. The B horizon is considered subsoil between the A and C horizons. Living organisms are fewer in the B horizon than the A horizon and are usually harder when dry and stickier when wet and may be of different color. The C horizon is the deepest of the three horizons and is parent material of the A and B horizons. It has less living matter and organic matter and is lighter in color than the A and B horizons. Soils can also have an O or organic horizon on the upper part of mineral soil dominated by fresh or partly decomposed organic matter.

Making of Soil

The accumulation of soil parent material follows the weathering of rocks which is a slow, gradual process. Decomposition of minerals usually proceeds long after the rock has disintegrated. Disintegration proceeds slowly and is influenced by climate. A mantle of weathered rock is called regolith. Plants soon grow in regolith. The addition of decayed plants, microorganisms and small animals and organic matter occur as soil horizons form.

Soils gain and lose organic matter, clay, and minerals, water, CO_2 and other elements depending upon leaching (eluviation) or illuviation. Illuviation is "washing out." Loss of some nutrient elements may occur due to plant growth.

Water must move through a soil profile before leaching and eluviation can happen. Leaching and eluviation affect soils in humid regions more than in arid regions. Soluble salts (sodium chloride), carbonates and elements such as calcium, sodium and potassium are slowly leached from the profiles of well-drained soils in humid areas. In arid areas soluble salts may become a problem for plant growth because of lack of nutrients, water and high pH (alkalinity) of the soil.

Simonson (1957) indicated five major factors in soil formation are climate, living organisms, parent rocks, topography, and time. These factors control weathering of rocks and the gains, losses and alterations throughout the regolith including the soil profile.

Soil Taxonomy

The thousands of soil types in the United States can be classified into about 49 great soil groups and double that for the world according to the 1938 classification (Soil Survey Staff 1975). For the world, Simonson (1957) listed six broad belts of Tundra, Podzolic, Chernozemic, Desertic, Latosolic, and soils of mountains. However, as a result of soil taxonomy approximation and testing over nearly two decades, the soil names have been changed. By 1971 about 9500 soil series were being used in the United States. The 1938 soil taxonomy classification system was revised in 1959 and is referenced in the U.S. Department of Agriculture Handbook No. 436 (1999). It provides a basic system of soil classification for making and interpreting soil surveys.

Orders

The nomenclature of soils is recognized by Orders followed by Suborders, Great Groups, Subgroups, Families, and finally Series. Names of Orders end in <u>sol</u> (*L. solum,* soil) with the connecting vowel <u>o</u> for Greek roots and <u>i</u> for other roots. Each name of an Order contains a formative element that begins with the vowel next proceeding the connecting vowel and ends with the last consonant proceeding the connecting vowel. For the Order name Entisol, the formative element is <u>ent</u>. For Aridisol it is <u>id</u>. These formative elements are used as endings for the names of Suborders, Great Groups, and Subgroups. Formative elements in names are given plus keys to identify them (Soil Survey Staff 1975).

Suborders

Suborders have two syllables. The first connotes diagnostic properties; the second in the formative element. The Suborder of Entisols that have an aquic moisture regime is called Aquents (*L. aqua,* water, plus <u>ent</u> from Entisol).

Great Groups

Names of Great Groups have three or four syllables and end with the name of a Suborder. For example, Fluvents that have a cryic temperature regime are called Cryofluvents (Gr. kryos, icy cold, plus fluvent). Fluvents that have a torric moisture regime are called Torrifluvents (*L. toridus,* hot and dry).

Subgroups

The name of a Subgroup consists of the name of a Great Group modified by one or more adjectives. The adjective typic is used for the Subgroup that is thought to typify the Great Group. The Torrifluvents that typify the central concept of the Great Group are called Typic Torrifluvents. Vertisols are called Vertic Torrifluvents if Torrifluvents have many but not all the properties diagnostic of Vertisols.

Families

Names of Families are polynomial. Each consists of the name of a Subgroup and adjectives, generally three or more to indicate the particle size class, the mineralogy classes, the temperature regime and sometimes depth of soil, consistence, moisture equivalent and other properties.

Series

Names of Series are usually abstract place names taken near where the Series was first recognized. Most names have been carried over from earlier classifications.

The names described have meaning and accurately define each soil considered relative to texture, temperature and moisture regime, sediments and other properties. Soils are a complicated and fascinating entity.

A complete guide to the identification of soils and the terms used to describe them are given in the 1999 U.S. Department of Agriculture Handbook No. 436. This information can be obtained online.

Also the terms to describe soil profiles given in the USDA Handbook No. 436 include description of soil profiles by capital letters, and by subdivisions if needed, or by an Arabic number after the capital letter such as O1, O2, A1, A2, A3, B1, B2, and B3. Other designations are indicated depending upon what is found in the soil horizon using small letters or roman numerals. Profile O is an organic matter layer.

Physical Properties of Soil

The flow of water and storage, movement of air, and the ability of soil to supply nutrients to plants are determined by soil particle size (Russell 1957). Inorganic soil particles occupy about one half of the total volume of most surface soils. Particles more than 2.0 mm in diameter are considered gravel or stones. Under 2.0 mm particles may be sand, 0.05 to 2 mm, silt, 0.002 to 0.05 mm, or clay which is less than 0.002 mm in diameter. The percentages of sand, silt and clay determine how soils react to water movement, chemical reactions and plant growth.

Soils with large amounts of gravel or sand have little or no plasticity and cannot retain large amounts of water or nutrients. Silt is between clay and sand and has greater chemical activity and plasticity and cohesion than coarser soil separates. The clay fraction is the one that controls most of the important properties of a soil. In temperate regions it is composed chiefly of secondary crystalline alumino-silicates which are plate-like in form. Hydrated sesquioxides of iron and aluminum are the main components of clay in the more weathered soil of the Tropics (Russell 1957).

Silica sheets are characteristic of the micas and the clay minerals. These sheets are bonded through common oxygen atoms with sheets of aluminum or magnesium octahedra.

Kaolinite, a major clay mineral especially in the Southeastern United States, consists of silica and alumina sheets in a 1:1 ratio. Two other important types of clay minerals - montmorillonite and illite - are composed of silica and alumina sheets bonded together in a 2:1 ratio (Russell 1957).

Ionic substitution of Al^{+++} for Si^{++++} and Mg^{++} or Fe^{++} for Al^{+++} are common in minerals of the 2:1 ratio. Such substitution gives clay mineral crystal a negative charge and causes clay particles to react with other charged particles, ions and dipolar molecules such as water, H^+, Ca^{++}, Mg^{++}, and K^+. This exchange process is important in soil management and plant nutrition.

Pore space as measured by bulk density occupies roughly one half of the total soil volume. It is occupied by water and soil air. It is useful in determining soil behavior (Russell 1957) such as rate of oxygen diffusion into and carbon dioxide out of the soil.

Dark-colored soils capture a much higher proportion of radiant energy than light-colored soils. This increases soil temperature in daylight hours and is reversed during the night. The soil surface is most affected by temperature fluctuations and affects water use and chemical reactions throughout the growing season.

Compaction of soils from mechanical manipulation and cultivation affect plant growth. Plants may be adversely affected by soil compaction. Soils high in organic matter are less affected by tillage than soils of high clay content. Soil crusting adversely affects plant growth and establishment also.

Of equal importance are the quantitative aspects of soil as a habitat for plant roots. It is necessary that nutrients, air and water are present in optimum concentrations for normal root development throughout the growing season (Russell 1957). Favorable air temperature, sunlight and length of growing season are also required.

Plant Needs from Soil

Adequate water is the medium soil nutrients are taken up by the plant for growth. Water is also the medium in which all physiological and biochemical reactions occur in the growing plant (Wadliegh 1957).

Growth of the annual seed plant is in four stages: seed germination, vegetative development, reproductive processes and the maturing of the seed. Absorption of water is the first step in germination of the seed that develops into a seedling plant. The second stage in germination is the use of stored nutrients by the developing seedling. As the plant develops vegetatively, it requires energy from sunlight with CO_2 to produce simple sugars (photosynthesis). CO_2 produced by respiration from microorganisms in the soil is an important source of CO_2. Favorable air temperature is also important for many plant biochemical processes for rapid plant development as well as CO_2 surrounding the plant.

The nitrogen in proteins is the key component. Proteinaceous substances are made up of mainly carbon, hydrogen, oxygen, nitrogen and sulfur. The first three elements come from sugars made by photosynthesis. The sulfur usually comes from soil. The nitrogen supply in the soil regulates the ability of the plant to make proteins that are necessary for new protoplasm formation. Inadequate nitrogen in soils limits plant growth (Wadliegh 1957).

Phosphorus is vital to processes in living cells - plant and animal. Combined with other organic substances, phosphorus provides biochemical energy transfer. When sugar is manufactured by photosynthesis or releases energy through respiration or synthesizes, proteins form new protoplasm, organic phosphorus compounds control the biochemical processes (Wadliegh 1957).

A third major element is potassium and is essential for plant growth. Potassium acts in enzymatic processes. Enzymes are protein-like substances that catalyze biochemical reactions. Many have a metallic makeup - like iron or copper - and these attached metals control the action of the enzyme. A plant cell may have hundreds of different enzymes active in different processes. Nitrogen, phosphorus and potassium are many times added as fertilizers to soils in certain ratios to promote growth of gardens, field and forage crops for improved yield and production.

Calcium is a highly essential plant nutrient and affects certain enzymes and is especially important in combining with protein to form calcium pectate - a cementing material between cells that holds them together. In deficient soils calcium can be applied as superphosphate and as liming materials. Lime also corrects soil acidity making them more neutral in pH and improving plant growth.

Other essential elements are magnesium for enzyme systems and chlorophyll production and sulfur for amino acid makeup. Although required at lower levels, iron, manganese, zinc, copper, boron, molybdenum and chlorine are also essential for plant growth.

Soil testing for nutrients or nutrient deficiency can be done by state or private testing laboratories for most soils from farm and urban areas. Soil testing is desired for optimum crop production to remove the uncertainty to apply or not to apply supplemental fertilizer.

Plants in poor health may be stunted or show discoloration that may be clues to nutrient deficiency. Climate may also influence expression of symptoms as well as poor soil drainage. Some plant diseases or insect damage could also cause poor growth. Excess salts in soil can cause reduced plant growth as well as drought. Causes other than nutrient deficiency for poor plant growth must be evaluated before taking action on soil testing.

Living Organisms in Soil

Living organisms in the soil enhance higher plant growth by improving fertility and soil tilth. Bacteria are the smallest and most numerous of the free-living organisms in soil (Clark 1957). Although they can be thousands in number per square inch, they weigh only a small fraction of the soil. They persist on waste organic matter. Heterotrophic bacteria derive both their cell carbon and energy from organic substances. Their ability to carry on life processes as oxidizing sugar to carbon dioxide and water is like that of higher animals and depends upon the energy stored in carbohydrates, fats and proteins.

A few bacteria possess pigments that enable them to trap the energy in light. They obtain their cell carbon directly from CO_2 or the atmosphere similar to green plants.

Autotrophic or chemosynthetic bacteria draw upon the atmosphere for carbon and obtain energy by oxidizing relatively simple chemical materials. In this group some oxidize carbon monoxide to carbon dioxide, sulfur to sulfates, hydrogen to water, ammonia to nitrous acid and nitrous acid to nitric acid.

Only a limited number of microorganisms can use nitrogen gas from the atmosphere. Soil bacteria that can do this are legume-nodule bacteria or rhizobia and can partnership with leguminous or other host plants. Both benefit from the nitrogen taken from the atmosphere. Nitrogen fixed by nodulated legumes can produce 50 to 150 pounds of nitrogen per acre.

Packaged inoculants containing nitrogen-fixing bacteria are available from seed stores for each specific legume crop to be grown.

Actinomycetes are microscopic organisms that resemble bacteria (Clark 1957). Actinomycetes, unlike bacteria, form long, threadlike, branched filaments and are sometimes called ray fungi. They are only about one-tenth to one-fifth as numerous as bacteria in most soils. Actinomycetes feed on organic residues and are important in the decomposition of and humidification of organic residues.

Fungi are less apparent in soils but total acre-weight is roughly equal to the combined weight of bacteria plus actinomycetes (Clark 1957). Many different fungi exist in soil. Fungi have no green pigment and many are parasitic on plants and animals. They are important in decay of organic residues but require oxygen to be active. Some fungi can colonize the surfaces of plant roots and can form fungus-root associations known as mycorrhiza. Such associations seem to benefit higher plants and increase the absorptive area of the roots manyfold.

Algae are the simplest forms of plant life and commonly possess photosynthetic pigments (Clark 1957). There are blue-green, green, red, and brown algae. They occur in greatest abundance as seaweed and pond scums but form only a minor fraction of soil microorganisms. Apart from their nitrogen fixation and their contribution to formation of organic soil, the role of algae in soil appears insignificant.

The bacteria, actinomycetes and fungi are agents of decay of organic matter. Together they are indispensable in the mineralization of plant and animal residue. The atmosphere above the surface of an acre contains about 20 tons of CO_2. The soil microflora living in an acre return about 20 tons of CO_2 each year to the atmosphere and are an essential part of the carbon cycle (Clark 1957). Knowing this, how does man control CO_2 levels on land and sea or does he need to? Does human activity really add much to the present enormous natural concentration of CO_2 in the environment?

Nitrogen stored in the soil is almost entirely organic nitrogen. Soil microflora are responsible for the release or mineralization of the organic nitrogen, making it available to plants. Nitrification occurs by oxidizing ammonia to nitrite by *Nitrosommas* bacteria and further to nitrate by other specialist bacteria such as *Nitrobacter*. Microbes also mineralize organic phosphorus in some soils and iron and manganese.

Other Soil Organisms

Other soil organisms include protozoa, nematodes, mites, and insects (Clark 1957). Protozoa and slime mold feed on bacteria. Protozoa are single-celled, microscopic and belong to the animal kingdom. They are slightly larger than bacteria. Slime molds are related to protozoa and feed on bacteria and form large jelly-like masses. They reproduce by spores.

Nematodes are non-segmented worms (Clark 1957). Most are microscopic but can attain a length of several inches or feet. They can parasitize plant roots. Total weight in an acre-foot of soil may be 50 pounds.

Mites and insects are present in field and forest soils (Clark 1957). Mites range in size from microscopic to barely visible in size. Their numbers can reach several billion in an acre of soil. They feed on other small animals, but most feed on plant wastes and fungi.

Many thousands of insect species occur in soils and attack living plants above ground, plants and plant litter at the soil surface and subsurface. Insect activity may affect the mineral-organic layer, porosity and water intake of soil and transport of soil.

Earthworms are important agents in mixing surface organic residues with underlying soil. They can burrow 6 feet deep but most intense activity is in the upper 6 inches of soil. They improve soil aeration and tilth.

Many mammals and birds may only have a slight effect on soil. Soil dwelling rodents, however, build nests and tunnels and transport large quantities of subsoil to the surface, which may or may not be desirable.

Toxic Elements in Soils

Toxic elements include selenium, arsenic, molybdenum, fluorine, manganese, iron, and aluminum (Bear 1957).

Selenium (Se) is a nonmetal and resembles sulfur in its chemical properties and occurs in North America where it has been derived from Cretaceous rocks. Lack of rainfall prevents the removal of selenium from some rocks. The primary problem is selenium accumulation in forage plants and use by livestock. Animals consuming selenium-accumulating plants may be harmed. Selenium toxicity can be overcome partly by feeding linseed oil meal supplement. The worst areas should be fenced off and posted.

Arsenic (As) is a gray, brittle, crystalline substance (Bear 1957). It can be toxic at 2 ppm (parts per million) to barley and alfalfa seedlings. Most tolerant fruits are apples, pears, grapes, and raspberries. Tolerant crops include rye, asparagus, cabbage, potatoes, and tomatoes. Bluegrass and orchardgrass are tolerant forage plants. Least tolerant crops include peaches, apricots, barley, wheat, peas, beans, alfalfa, clovers, vetch, and sudangrass. Other crops are intermediate in tolerance.

Treatment for arsenic toxicity consists of 1 to 2 tons of iron sulfate per acre or triple-superphosphate. Zinc sulfate at 5 to 10 pounds a tree is used on peach trees. Arsenic toxicity is recognized by slow, stunted growth and late maturity.

Molybdenum (a catalyst in enzyme systems in plants) above 10 ppm in dry green forage can be toxic to animals. To overcome molybdenum toxicity, grow forage crops on the land for hay until high concentrations are removed from the soil and hauled away with the hay.

Fluorine is usually 2 to 20 ppm in dry weight plants. Injury to plants occurs at 50 ppm and plants may die at 500 ppm. Soils normally contain 100 to 300 ppm. A person takes in about 0.5 to 1.0 milligrams (mg) of fluorine daily and is added in some areas to the drinking water to prevent tooth decay.

Boron is needed by plants at about 25 to 75 ppm in soils but can become toxic in excess. For sensitive crops the boron limit is 0.66 ppm, for semi-intolerant crops the limit is set at 1.33 ppm and for tolerant crops at 2 ppm.

Manganese and iron are paired because they may interfere with each other as catalysts in enzyme systems that are needed for life. One that is present in excess may substitute for the other to the extent that plants are injured. Therefore it must be determined what ratio of each is best for the plant in question. Liming of soils can raise the pH value to around 6. Liming the soil causes a lowering in the solubility of manganese and helps bring the iron and manganese into better balance for the plant.

Aluminum is toxic to plants in acid soils and toxicity can be greatly reduced by liming acid soils. Other toxic elements can include barium, nickel, copper, zinc, and lead. Almost any soil element is toxic to plants in extremely high concentrations. Application of the wrong or improper rates of fertilizer, insecticides, fungicides or herbicides can be toxic to crop plants.

Soil Management

Most farmers prepare their soil for seeding and weed and pest control by tillage (Horowitz et al. 2010). The purpose of tillage is to change the structure of the soil to kill weeds and manage crop residues (Raney and Zingg 1957). It also facilitates the intake, storage, and transport of water for preparation of a good seedbed and environment for plant roots.

Planting, cultivating, and harvesting operations may adversely affect soil structure. A desirable soil structure is one in which large stable pores extend from the surface to the water table or drains (Raney and Zingg 1957). Such pores ensure rapid infiltration and drainage of water and allow aeration of the subsoil. Small pores in soil store water. The large and small pores allow adequate storage, intake, and transport of water. Theoretically, the best sizes of aggregates range from 1 to 5 mm in diameter (Raney and Zingg 1957).

The old standard method of breaking land consists of plowing a furrow slice cut loose by the plowshare. It should be done when soil moisture is correct and other implements such as disks and harrows are used to further break up sod or soil clods for a good seedbed. However, Horowitz et al. (2010) showed that about 35 percent of the U.S. cropland planted to eight major crops or 88 million acres had no tillage operations in 2009. This included barley, corn, cotton, oats, rice, sorghum, soybeans, and wheat. No-tillage means planting crop seed directly into crop and weed residues without further soil cultivation (Throckmorton 1986). No-till may include planting, fertilizer application and weed and insect control in one pass over the field. Guidance systems and seed and chemical applications are computer controlled and monitored. Reduced tillage operations save time, fuel, labor, and machinery costs and reduce soil erosion (wind and water).

Other surface tillage methods include chisel plows, disking, V-sweep plows, tandem disk harrows, and field cultivators (Throckmorton 1986). Special seedbed conditioners, roller harrows, tillage-seeders, and drills are available for soil preparation and planting. A no-till planting system is characterized by no-overall-tillage or stalk-reduction. A narrow-tillage strip is the only tillage performed. Some designs use a flat coulter preceding the planter opener. Griffith et al. (1987) indicated that the no-tillage system had more soil organic matter, decreased evaporation loss of moisture, increased rainfall infiltration, improved soil aggregation at 0.5 mm and at 5-15 mm and more undisturbed large soil pores and reduced soil erosion compared to conventional farming operations.

Erosion Control

As just discussed, no-till farming significantly reduces wind and water erosion of cultivated land. Wind and rainfall intensity determines the degree of soil erosion. Soil erosion is the physical process of wind or water picking up and moving soil particles (Nebel 1987). A cover of vegetation is the first defense against erosion and runoff (Blakely et al. 1957). A soil protected by the proper amount of sod or forest litter is unlikely to erode regardless of wind or rainfall intensity. A protected soil also absorbs rainfall, and good structure permits free movement and water storage.

Cropland gives intermediate soil erosion protection between plowed areas versus sod or forest litter. Hay crops give good soil protection. Crop rotation of fallow wheat and kefir leaves soil exposed for most of a year, but wheat planted the year followed by alfalfa-bromegrass mix for three years gives good protection most of the time. Contour tillage, stripcropping, and terracing have been used to reduce soil erosion in the past.

The most erodible soil particles by wind are about 0.1 millimeter in diameter (Chepil 1957). A rough ground surface is more resistant to wind erosion than a smooth surface. A rough surface slows down wind velocity and tends to trap dislodged particles. Vegetation and vegetation residues protect soil similar to rough surfaces. Soil texture also influences movement. The coarsest and often the finest textured soils are more erodible than medium-textured soils. Coarse-textured soils lack sufficient silt and clay to bind the erodible sand grains. The fine-textured soils have too much clay, which may cause soil clods to disintegrate into a finely granulated erodible situation. Soil particles greater than one millimeter in diameter, whether single grains or aggregates, are highly resistant to wind erosion.

To control wind erosion, protect soil with vegetation and crop residues. Roughen the surface to slow wind erosion of cropland, produce stable aggregates large enough to resist movement, and place barriers such as crop strips, ridges or shelterbelts to trap the wind and prevent soil from drifting.

Soil Amendments

McVickar (1970) indicated there are 16 elements known to be necessary for a plant to grow normally. They are carbon, hydrogen, oxygen, nitrogen, phosphorus, calcium, magnesium, sulfur, manganese, boron, copper, zinc, iron, molybdenum and chlorine. Brady (1990) also lists cobalt as a required micronutrient in soils necessary for plant growth. Nine of the seventeen plant food elements are required in relatively large amounts. They are hydrogen, oxygen, carbon, nitrogen, phosphorus, potassium, calcium, magnesium, and sulfur.

Plants get hydrogen and oxygen from water and carbon dioxide (carbon and oxygen) from the air. Under most conditions water and air provide adequate amounts of hydrogen, oxygen, and carbon for plant life. Carbon dioxide is necessary for photosynthesis. Carbon dioxide is released when the plant residues decay.

Nitrogen

The air around us is composed largely of nitrogen, an inert gas. In order for plants to capture this nitrogen, certain legumes and other non-legume plants that fix nitrogen can make nitrogen available to plants as a result of inoculation or natural fixation by soil bacteria. Nitrogen can also be made available to plants from decay of plant residues by soil microorganisms or by electrical discharge in storms which causes nitrogen and oxygen to combine. The oxides formed dissolve in rain to form nitrous or nitric acids in soil which combine with mineral salts to form nitrites and nitrates. Nitrites and ammonia in soil from animal excretion and/or decay of organic matter can be converted to nitrate by nitrification. Nitrogen is fixed industrially by combining it with hydrogen in the Haber process and is added to soil as fertilizer (Bailey 2006).

Crop and native plants use large amounts of nitrogen. Most nitrogen is used as nitrate (NO_3^-) ions or as ammonium (NH_4^+) or nitrite (NO_2^-). Deficiency leads to spindly growth and yellowing of the leaves. Nitrogen. as stated above, is replenished by natural processes or application of compounds such as urea, ammonium nitrate, ammonium sulfate, and nitrochalk. Many high-yielding crops depend upon high levels of nitrogen (McVickar 1970). For example, use of nitrogen in Iowa, U.S.A. in many fields is currently about 150 kg/ha (135 lb/A) due to increased mechanization and management of soil, crop residue, and chemical and genetic (hybrids) inputs (Smith et al. 2015).

McVickar (1970) indicated the principal inorganic nitrogen-carrying materials include sulfate of ammonia, anhydrous and liquid ammonia, nitrate of soda, ammonium nitrate, ammonium phosphates, calcium nitrate, nitric phosphates, and nitrate of potash. Anhydrous ammonia is popular and is applied directly to the soil or through irrigation water. Application directly to soil requires special safety equipment and must be accurately metered out of the ammonia tank to a depth of 4 to 6 inches in the soil. Anhydrous ammonia can also be dissolved in water and is called aqua ammonia. To avoid loss, nitrogen aqua ammonia should be injected below the soil surface or below the surface of water when applied.

Synthetic organic nitrogen materials include urea, urea-phosphates, urea-sulfur, urea-formaldehyde, and calcium cy-anamide. All inorganic and organic nitrogen fertilizers vary in nitrogen content but are used based on cost, availability and specific use (McVickar 1970). Natural organic nitrogen materials are derived from plants and animal wastes. They are largely insoluble in water in contrast to the inorganic nitrogen materials that are readily soluble. Organic materials used in fertilizers include cottonseed meal, tankage, linseed meal, castor pomace, bone meal, tobacco byproducts and guano. They were readily used 100 years ago, but have largely been replaced by the synthetic inorganic or organic fertilizers. Organic materials decompose slowly in soil; consequently, the nitrogen slowly becomes available.

Nitrogen fertilizer total N composition is usually about 20% by weight but may vary from 3 to 45% depending upon its chemistry and manufacturing process. Total fertilizer use has increased markedly in the last 100 years, but high costs have slowed use.

Phosphorus

Phosphatic fertilizer materials are derived from three main sources: rock deposits and lesser amounts from iron ore and bones.

Phosphatic fertilizer content is usually expressed as phosphoric acid (P_2O_5). Pure phosphorus cannot be used as a fertilizer as such but must be combined with other elements as carriers before it can be safely handled and used as a plant nutrient.

All living cells contain phosphorus and plants and animals become stunted, diseased and spindly if deficient. The average percentage of phosphorus in surface soils is only half that of nitrogen and a fraction of that of potassium. Fertile soils contain no more than 1500 pounds of phosphorus in the plow layer (McVickar 1970).

Phosphorus is absorbed by the plant as a phosphate ion as H_2PO_4, HPO_4 or PO_4 depending upon soil acidity. The quantities of these forms in soil solution are usually less than one-half part phosphorus per million.

Phosphate fertilizers include superphosphate (16 to 20% available phosphoric acid) or concentrated superphosphate (40 to 50% available phosphoric acid), ammonium phosphates. nitric phosphates, liquid phosphoric acid and many others (McVickar 1970).

Potassium

Like nitrogen and phosphorus, potassium must be combined with other elements before it can be used as a fertilizer (McVickar 1970). Most potassium in soil is relatively insoluble even though the soil may contain as much as 40,000 pounds per acre. The potassium in fertilizer is expressed as potash (K_2O). Potash salts are mined from mostly in the western U.S. The potash in commercial fertilizers are muriate of potash (KCl), sulfate of potash (K_2SO_4), and nitrate of potash (KNO_3).

Plants need large amounts of potassium; and it is supplied to plants by natural sources in the soil, fertilizers, manures and mulches (Reitemeier 1957).

Some loams and clays containing abundance of illitic clay minerals or unweathered primary potassium minerals can be expected to supply adequate potassium for a cropping system. Early in the cropping system younger soils may not need potash fertilizer. Available soil potassium can be determined by a soil test or by a plant tissue test (Reitemeier 1957).

Potassium can be supplied to soils by fertilizers of muriate of potash (potassium chloride) (KCl), sulfate of potash (potassium sulfate) (K_2SO_4), manure salts and kainit, nitrate of potash (potassium nitrate) (KNO_3) and others that contain high amounts of potash.

Nitrogen, phosphate and potassium are the three major plant nutrients that are required in relatively high amounts. Some fertilizers, depending on crop needs, are made up of all three nutrients or in certain combinations of nitrogen, phosphoric acid, or potash. An example of a nitrogen fertilizer would be ammonium nitrate containing about 34% nitrogen. Common ingredients would be nitrogen at 16% and phosphoric acid at 20% in ammonium phosphate sulfate. A blend of nitric phosphate and ammonium phosphate could contain 16% each of total nitrogen available P_2O_5 and available K_2O (McVickar 1970).

Secondary and Trace Elements

As previously mentioned, the secondary elements are sulfur, calcium, and magnesium. The trace elements are manganese, boron, copper, zinc, iron, molybdenum, chlorine, and cobalt.

Sulfur

Sulfur is an essential plant and animal element, and sulfuric acid is used to convert insoluble phosphate to superphosphates and phosphoric acid (McVickar 1970). Sulfur is essential to amino acid formation and plant metabolism. Sulfur is used to control various pathogens and pests. Sulfur is absorbed from soil as the sulfate ion (SO_4^{2-}).

Calcium

Certain fertilizers have large amounts of calcium. Superphosphate is about 20% calcium. Other fertilizers like nitric phosphate, calcium cyanide, calcium nitrate, and rock phosphate are important sources of calcium. As agriculture soils become older, the need for calcium intensifies. Calcium is essential to plant growth due to its role in the formation of calcium pectin and cell rigidity.

Magnesium

Magnesium can be supplied in dolomite limestone, magnesium sulfate, sulfate of potash magnesia, and magnesium oxide. Magnesium is essential in chlorophyll production - the green color in plants. When deficient in crops fed to livestock, the animals also suffer.

Manganese

Manganese is essential to plants and animals in extremely small amounts and functions in certain enzyme systems and with other elements. Manganese sulfate is water-soluble and can be used for soil application, fertilizer mixes, and foliar feeding when deficient. Manganese is essential for seed development. It can apparently be toxic at high levels with low levels of zinc and copper.

Some soils may have as much as 3000 parts per million, but most are unavailable for plants (Schulte and Kelling 1999). Manganese reactions in soil are complex. Manganese toxicity is common in acid soils below pH 5.5 and may be deficient in soils above pH 6.5. Liming is used to prevent toxicity in acid soils.

Manganese fertilizer recommendations of MnO and $MnSO_4$ are recommended for soil and foliar application, respectively, for crops from 3 to 5 pounds per acre for soils and 0.75 to 1 for foliar application (Schulte and Kelling 1999). Mn chelate can be applied to soil or foliar at lower rates. In soil, manganese occurs as exchangeable manganese, manganese oxide. organic manganese, and a combination of ferro-magnesium silicate minerals. The manganese ion (Mn^{++}) is similar to the size of magnesium (Mg^{++}) and ferrous ion (Fe^{++}) and can be substituted for these elements in silicate minerals and iron oxides.

Boron

Boron is a metalloid element found in very low concentrations in plant tissues. Its role is uncertain, but it is associated with cell membrane functions and in cell extensions (Bailey 2006).

The use of boron dates prior to 1556 when Arabians and Persians used "Borak" - a white powder - in glazes and was used in agriculture about 400 years later as a plant food (McVickar 1970).

Copper

Copper is an essential in trace amounts for growth. It is found in the enzyme cytochrome c oxidase which catalyzes the transfer of electrons from cytochrome a3 to oxygen in respiration. Copper deficiency leads to chlorosis so it is associated with chlorophyll (Bailey 2006). Copper can be added to common fertilizers as copper oxide (McVickar 1970).

Zinc

Zinc is a metal element needed in trace amounts. Zinc ions are required as cofactors to certain enzymes (Bailey 2006). Zinc used in agriculture is in the form of zinc sulfate, zinc oxide, and zinc nitrate (McVickar 1970).

Others like zinc chloride, zinc carbonate and metallic zinc have been used. Large quantities of dolomitic limestone containing zinc have been used. Chelating materials have also been used (an organic compound combined with zinc). Most are restricted to spray (foliar) treatment.

Iron

Iron is abundant in soil, but the solubility of iron is governed by soil reaction. Acid-forming fertilizers, elemental sulfur, or ammonium sulfate increase the acidity which, in turn, increases iron availability. Liberal application of organic matter can also increase iron availability. Over-liming may induce iron chlorosis (McVickar 1970).

Iron is an important constituent of the cytochromes, ferredoxin, and certain enzymes.

Deficiency leads to iron chlorosis (green plants becoming unhealthy, pale or yellow in color) (Bailey 2006).

Molybdenum

Molybdenum in plants is involved in the nitrate reductase enzyme system and nitrogen fixation process (Bailey 2006). Soil deficiencies are few.

Chlorine

Chlorine is essential in small amounts. It is associated in the regulation of osmotic and ionic balance in plants and may be involved in the light reaction of photosynthesis (Bailey 2006). Chlorine deficiency leads to wilting and later chlorosis.

Cobalt

Essential for nitrogen fixation and found in vitamin B_{12} (Brady 1990).

Soil Problems

Soil problems include compaction, disease problems, crusting, drainage, soil life, salinity, erosion, infiltration, lack of organic matter, and soil pH (Anonymous 2011).

Compaction

To prevent compaction, avoid working soil when wet and reduce tillage, add cover crops and use crop rotation, add animal manures and use non-compacting tillage.

Crop Disease

Check for pathogens and pests, correct soil nutrients and pH levels, and follow practices listed for compaction.

Crusting

Increase organic residues, reduce tillage depth, use animal manures, and add cover crops. For sodium problems, apply gypsum and flush with irrigation water.

Drainage

Subsoil to break up tillage pan, add a drainage system.

Soil Life

Includes reduced earthworm and microbial activity caused by low organic matter, excessive fertilization or pesticides, excessive tillage or poor aeration. Soil life can be improved by conservation tillage, cover crops, and crop rotation.

Salinity

Salinity problems can be observed by a white crust on soil and electric conductivity tests. The excessive salts can be leached by irrigation water, removed by deep rooted salt-tolerant crops, and improved drainage.

Erosion

Loss of top soil and creation of rills and gullies are indicative of soil loss. Reduced tillage, animal manure, cover crops, crop rotation, crop strips/windbreaks, and leaving the soil surface rough and/or with crop residue are some major ways to prevent erosion.

Infiltration

Poor infiltration of water results from lack of cover and organic residues, low organic matter, poor aggregation of soil particles, compaction, and excessive tillage. To alleviate infiltration problems, add organic residues, animal manure, cover crops, rotate crops, subsoil, and use tillage that preserves soil structure. For sodium problems, apply gypsum and flush with irrigation water.

Organic Matter/Residues

For lack of organic matter, diversify or increase crop rotations, add animal manure, use cover crops and high residue crops, and reduce tillage.

Soil pH

Use ammonium fertilizers and no lime, but use lime for low pH and improve drainage. A neutral soil pH is best since most soil nutrients are readily available and in favorable balance for plant growth.

Summary

This chapter by no means captures all the detail on soils but is merely a snapshot of their importance. Good soil resources and proper management is basic to any prosperous and independent nation for food and well-being. A prosperous agriculture frees up society so other important activities and commerce can take place. The United States is blessed with adequate and many different types of soil for crop and animal production. Unfortunately, much of the productive land is lost each year to roadways, railways, residential areas, airports, and commercial installations. This means we will be forced to make better use of the remaining agriculture land to feed an exploding population.

The making of new soil occurs but it is a slow and gradual process taking many generations. Soil suitability for crop or animal production must be combined with favorable climate and rainfall. Soil must be a favorable mix of microbes, organic matter, small animals, nutrients, soil particles, water, CO_2, oxygen, minerals, and environmental factors to be useful.

Thousands of soil types occur in the United States and world. Many have been classified and surveyed giving clues to their management needs.

In addition to sunlight, water, a favorable temperature, and carbon dioxide to produce simple sugars (photosynthesis), plants need nitrogen to produce proteins. Sulfur is also essential for plant growth being found in amino acids, iron-sulfur proteins and is necessary in the electron transport system. It is also found in various secondary metabolites and is supplied by the soil. In addition to nitrogen, phosphorus and potassium are vital to plants and animal life. Other essential nutrients include calcium, magnesium and lower levels of iron, manganese, zinc, copper, boron, molybdenum, chlorine, and cobalt.

If plants or animals show deficiency symptoms, soil testing for nutrients can be done at state or private laboratories. Plants in poor health may be stunted or show discoloration.

Organisms living in the soil are also essential by improving fertility or soil tilth. They include bacteria, actinomycetes, and fungi. They are responsible in the mineralization of plant and animal residues. Other organisms include protozoa, nematodes, mites and insects, and earthworms.

Toxic elements such as selenium, arsenic, molybdenum, fluorine, manganese, iron, and aluminum can be toxic in soil if concentrations are too high. Methods to overcome toxicity are available.

Minimum or no-tillage operations on U.S. farmlands are becoming very popular. Over 35% of croplands are treated by this method. The advantages include reduced soil erosion and tillage operations and reduced operation time, fuel, labor, and machinery costs.

Soil management problems such as compaction, disease, crusting, drainage, salinity, erosion, soil pH, and others are discussed.

Nutrient needs and fertilizer possibilities to remedy plant-soil nutrient deficiencies are also discussed.

Chapter 2. Literature Cited

Wiley HW (1895). *Soil Ferments Important in Agriculture: Vitality of Soil.* Yearbook of the United States Department of Agriculture. 1895. Government Printing Office. Washington D.C. Pp 69-102.

Gardner FD (2001). *Traditional American Farming Techniques* (originally published in 1916 by LT Meyers as *Successful Farming*) First Lyon Press. Guilford, Connecticut. 1088 p.

Simonson RW (1957). What are Soils, Pages 17-31 in Stefferud A ed. Soil. The 1957 Yearbook of Agriculture. The United States Department of Agriculture. Superintendent of Documents. U.S .Government Printing Office. Washington D.C.

Soil Survey Staff (1975) *Soil Taxonomy: A Basic System of Soil Classification for Making and Interpreting Soil Surveys.* Soil Conservation Service (now Natural Resources Conservation Service) U.S. Department of Agriculture. Agriculture Handbook No. 436. Superintendent of Documents. U.S. Government Printing Office. Washington D.C. 754 p.

Russell MB (1957). Physical Properties. Pages 31-38 in Stefferud A ed. Soil. The 1957 Yearbook of the Agriculture, U.S. Department of Agriculture. Superintendent of Documents. U.S. Government Printing Office. Washington D.C.

Wadleigh CH (1957). Growth of Plants. Pages 38-49 in Stefferud A ed. Soil. The 1957 Yearbook of the Agriculture, U.S. Department of Agriculture. Superintendent of Documents. U.S. Government Printing Office. Washington D.C.

Clark FE (1975). Living Organisms in the Soil. Pages 157-166 in Stefferud A ed. Soil. The Yearbook of the Agriculture, U.S. Department of Agriculture. Superintendent of Documents. U.S. Government Printing Office. Washington D.C.

Bear FE (1957). Toxic Elements in Soils. Pages 165-171 in Stefferud A ed. Soil. The Yearbook of the Agriculture, U.S. Department of Agriculture. Superintendent of Documents. U.S. Government Printing Office. Washington, D.C.

Horowitz J, Robert E, Kohei U (2010). "No-Till" Farming is a Growing Practice. Economic Research Service. U.S. Department of Agriculture, Economic Information Bulletin No. 70. Unknown p.

Raney WA, Zingg AW (1957). Principles of Tillage. Pages 277-281 in Stefferud A ed. Soil. The Yearbook of the Agriculture, U.S. Department of Agriculture. Superintendent of Documents, U.S. Government Printing Office. Washington D.C.

Throckmorton RI (1986). Tillage and Planting Equipment for Reduced Tillage. Pages 59-91 in No-Tillage and Surface-Tillage Agriculture: The Tillage Revolution. Sprague, MA and Triplett GB, eds. John Wiley & Sons. New York.

Griffith DR, Mannering JV, Box JE (1986) Soil and Moisture Management with Reduced Tillage. Pages 19-57 in No-Tillage and Surface-Tillage Agriculture: The Tillage Revolution. Sprague MA and Triplett GB, eds. John Wiley & Sons. New York.

Blakely BD, Coyle JJ, Steele JG (1957). Erosion on Cultivated Land. Pages 290-307 in Stefferud A ed. Soil. The 1957 Yearbook of the Agriculture, U.S. Department of Agriculture. Superintendent of Documents. U.S. Government Printing Office. Washington D.C.

Nebel BJ (1987) *Environmental Science: The Way the World Works.* 2nd edition. Prentice-Hall, Inc., Englewood Cliffs, New Jersey. 671 p.

McVickar MH (1970). *Using Commercial Fertilizers: Commercial Fertilizers and Crop Production.* 3rd edition. The Interstate Printers and Publishers, Inc. Danville, Illinois. 352 p.

Bailey J (2006). *Collins Dictionary of Botany.* Harper Collins Publishers, Westerhill Road, Bishopbriggs, Glasgow. 504 p.

Smith S, Cooper M, Gogerty J, Loffler C, Borcherding D, Wright K (2015). Maize. Pages 125-177 in Smith S, Diers B, Specht J, Carver B, eds Yield Gains in Major U.S. Field Crops. CSSA Special Publication 33. Crop Science Society of America, Madison, Wisconsin.

Reitemeier RF (1957). Soil Potassium and Fertility. Pages 101-106 in Stefferud A ed. Soil. The 1957 Yearbook of the Agriculture, U.S. Department of Agriculture. Superintendent of Documents. U.S. Government Printing Office. Washington D.C.

Stout DR, Thomson CM (1957). Trace Elements. Pages 139-150 in Stefferud A ed. Soil. The 1957 Yearbook of the Agriculture, U.S. Department of Agriculture. Superintendent of Documents. U.S. Government Printing Office. Washington D.C.

Schulte EE, Kelling KA (1999). Understanding Plant Nutrients: Soil and Applied Manganese. A2526. University of Wisconsin-Extension, Madison, Wisconsin. RP-08-2004-(SR 07/99) Unknown p.

Anonymous (2011). Soil Quality for Environmental Health. NRCS East National Technology Support Center and the University of Illinois, Urbana-Champaign, Illinois. Retrieved from soilquality.org/dev/page.cfm?pageid=secondary_functions... Accessed 27 July 2015.

Brady NC (1990). *The Nature and Properties of Soils*. 10th edition. MacMillian Publishing Company. New York. 621 p.

WATER

Introduction

One could write the story of humanity based on its reliance on water (Frank 1955). Water has shaped the growth of man since ancient times since all life depends upon water. When not available, man has devised methods in deficient areas to divert streams and rivers, dig wells, and/or make channels or pipe systems to meet water needs. In areas of abundant or adequate water and favorable weather and soils, the human population has done well.

As indicated in Chapter 2 on soils, plants depend upon water for survival and most chemical reactions in the plant and soil take place in water solution. The same can be said for animals that feed on plants or other organisms. As learned in Chapter 2, all mineral nutrients, except for nitrogen compounds derived from the atmosphere, are originally from rocks along with non-nutritional chemicals such as silica and aluminum (Nebel 1987). As weathering occurs, nutrient ions are released into water solution and can be used by plants and animals.

Water and Nutrient Holding Capacity

The ability of soil to loosely bind and hold nutrient ions from leaching is called the ion-exchange capacity. This is important in agriculture because nutrients used must be routinely replaced by natural supplies or replaced by addition of fertilizer as reported in Chapter 2. In addition to nutrient-holding capacity, soils also have a related capacity known as the water-holding capacity. Growing plants depend on water for photosynthesis (light plus water plus carbon dioxide produce carbohydrates).

Water Transpiration in Plants

During growth and photosynthesis, plants transpire (loss of water vapor) large amounts of water to the atmosphere, and water lost from the top of the plant must be continually replaced by absorption through the roots from water reserves in the soil. Without adequate water, plants may wilt and die. Forest trees transpire large amounts of water. Desert plants, conversely, may lose relatively little water and can survive in arid environments.

Evaporation of water through openings in the leaves (stomata) produces the energy gradient that is the principal cause of water movement into and through plants and controls the rate of absorption and ascent of the sap (Robbins et al. 1957; Galston et al. 1980; Kramer 1983). Transpiration may also cool leaves (Kramer 1983). It has been estimated that 1 acre of corn may transpire as much as 0.5 million gallons of water in one growing season (Nebel 1987).

Pores in the leaf surface, called stomata, which allow escape and evaporation of water from the plant, also allow exchange of carbon dioxide (CO_2) and oxygen (O_2) for photosynthesis (Nebel 1987).

Hydrological Cycle

The water cycle basically consists of an alteration of evaporation and condensation (Nebel 1987). Water molecules enter the air by evaporation and transpiration of plants. Once in the atmosphere, water molecules condense and return to the earth as precipitation. Some water will be absorbed by the soil and used by plants and animals. Some water will be held in and on the soil, runoff in surface water or become groundwater, but eventually it will all re-evaporate and complete the cycle.

Water Vapor

Spencer (2003) stated that although water is an exceptionally important component of the atmosphere, it is present in relatively small quantities (0 to 4%) by volume in the homosphere. The amount of water that evaporates from the earth's surface is about 425 cm (14 feet) annually. Atmospheric water plays an important role in weather and climate.

Nebel (1987) indicated that since about 70% of the earth's surface is ocean that a large amount of water evaporation occurs from its surface. Lakes, rivers, moist soil, and wet surfaces also contribute to evaporation. Although the atmosphere contains dry gases, pollutants, particulates, and greenhouse gases, it is predominantly nitrogen (78.1%) and oxygen (20.9%).

Temperature Zones

Temperature normally decreases at higher elevations in the atmosphere. The rate of decrease in temperature is 6.5°C per kilometer (3.5°F per 1000 feet) rising elevation. At 10 km (6 miles) the temperature decreases to about -60°C (-108°F). It remains stable between 10 to 20 km (6 to 12 miles) but begins to rise at higher elevations. Meteorologists, based on changes in temperature, subdivide the atmosphere into zones so the composition and concentration of certain atmospheric gases vary with altitude.

Composition of the Lower Atmosphere

As indicated in the troposphere (0 to 10 km high), nitrogen and oxygen are by far the most abundant gases with small quantities of water vapor and other particles and gases. This homogenous mixture is called the homosphere. Above the homosphere turbulence is greatly reduced and concentrations of oxygen and nitrogen form thick, poorly mixed layers. This part of the atmosphere (0 to 47 km) is called the heterosphere and contains concentrations of electrons and ionized particles of oxygen and nitrogen. The highest part of the atmosphere is extremely rarefied and contains hydrogen and helium that escape from earth and incoming cosmic rays, meteorites, and particles coming from the sun.

Humidity

Humidity is the quantity of water held in a given volume of air (Spencer 2003). <u>Absolute humidity</u> is the quantity of water vapor contained per unit of volume of air or the number of grams of water vapor contained in a cubic meter of air. <u>Relative humidity</u> is the ratio of the amount of water vapor in a volume of air to the amount of water vapor that could be contained in the air at a given temperature without having some change state and become liquid. Air is <u>saturated</u> with water vapor when the quantity of water is sufficient for some to condense to a liquid. The maximum quantity of water vapor that a given volume of air can hold depends upon temperature. Warm air holds more water vapor than cold air. If air temperature is above freezing, water vapor in saturated air condenses as liquid droplets. If air temperature is below freezing, water vapor may crystallize, forming snowflakes or ice. Precipitation consists of all types of falling water or ice. Most precipitation falls from particular types of clouds. Fog and droplets that form near ground level create ground-level clouds. The dew point is when the temperature cools and the water vapor begins to condense into dew.

Cloud Formation and Precipitation

Nothing is more important in dryland farming than receiving adequate precipitation at the right time to enhance crop, grazing land, and forestry production. Clouds are classified according to their shape, composition, and altitude (Spencer 2003). Most are stratus, cirrus, or cumulus clouds. Stratus are layered and generally below 2000 m (6560 feet) and produce rain or drizzle. At higher altitudes they are altostratus clouds. Cirrus clouds produce ice crystals blown by winds at higher altitudes over 6 km (20,000 feet). Cumulus clouds are puffy and resemble cotton balls and have a flat base that is often dark. Cumulonimbus clouds are huge, sometimes in anvil shape, and rise high into the atmosphere and produce thunderstorms.

The amount of precipitation occurring in an area is a primary factor in determining the type of ecosystem (Nebel 1987). The distribution of precipitation over the earth is basically dependent upon patterns of heating and cooling of the earth's atmosphere. It ranges from near 0 to 3 m (120 inches) per year. In equatorial regions high rainfall occurs from solar heating and rising currents of air. Air expands with increasing temperature and becomes less dense and rises. Rising air cools and the cool air cannot contain its vapor and rainfall results. Subtropical regions (25° to 35° north and south of the equator) are typified by deserts. Deserts have low rainfall because the rising air has dropped its moisture over equatorial areas before descending over subtropical regions.

Much of agriculture in Sub-Saharan Africa, Latin America, the Near East, and North Africa and in East and South Asia depend on rainfall for crop production. Rainfall can be inconsistent and crop production uncertain also due to degraded soils, high levels of evaporation, droughts, floods, and lack of good water management. Rainfed crops, grazing land, and forests the world over depend on natural rainfall events which sometimes fail.

Atmosphere Problems

As expressed in Chapter 1, there is a concern about the effect of chlorine on the ozone layer in the stratosphere (Spencer 2003). Ozone in the stratosphere (layer above the troposphere and below the mesosphere about 10 to 46 km altitude) serves as a shield to protect living organisms on earth from ultraviolet (UV) radiation. Chlorine is contained in chlorofluorocarbons (CFCs like Freon). CFCs break down ozone, and attempts have been made to limit their use in air conditioners, refrigerators, aerosol cans, and other sources.

Other atmospheric pollutants toxic to humans that may occur in air or water include carbon monoxide, nitrogen oxides, ozone, benzene and other organic compounds, lead, sulfur dioxide, carbon soot, and radioactive particles.

Acid rain (carbonic acid H_2CO_3) from carbon dioxide and water vapor results in rainwater of the pH of about 5.5. All forms of precipitation (rain, snow, ice, and fog) can be acidic. Other acids can form from sulfur dioxide (SO_2) and nitrogen oxides (NOx) as a result of volcanoes and burning of sulfur-contained coal. Vehicles also produce nitrogen oxides. Acid rainfall can disrupt the well-being of many plants and animals when severe. The Clean Air Act in the United States helps control acid rain pollution.

Irrigation Methods

Many locations in the world have favorable soils and climate but lack adequate water to grow crops. In such cases supplemental water by irrigation from water wells, rivers, streams, and lakes may be possible.

There are mainly five types of irrigation including surface irrigation where water is moved over the land by simple gravity to infiltrate soil (Anonymous 2009). It can be subdivided into fur-loughs, border strips, or basin irrigation. A second type is localized irrigation where water is distributed under low pressure by a pipe system. It can be drip irrigation, spray or micro-sprinkler, or bubbler irrigation. Third is sprinkler irrigation or overhead irrigation where water is piped to one or more central locations within the field and distributed (solid-set irrigation). In contrast, high-pressure sprinklers that rotate are called rotors and are driven by a ball drive or impaction mechanism to rotate in a field or partial circle. Guns are similar to rotors except they operate at a very high pressure. A fourth type of irrigation is subirrigation and involves applying water beneath the soil surface to create or maintain an artificial water table at some predetermined depth. The fifth method is manual irrigation using buckets or watering cans.

For home and garden use, sprinklers that deliver large drops versus a spray of fine mist conserve water from loss to evaporation (Anonymous 2015a). Soaker hoses that sweat water along their entire length also thoroughly wet soils and conserve water. Drip irrigation is the most efficient way to irrigate plants and soils, and it uses only a fraction of water compared to overhead spraying devices. Traditional automatic pop-up spray systems can be adjusted to spray full circle, half circle, or quarter circle. They are less efficient than rotor heads or drip systems because they spray water faster than some soils can absorb it. Rotor systems apply water slower than spray heads allowing the soil to absorb moisture more efficiently.

Other devices and techniques to conserve water in home and garden use are use of filters, pressure regulators, rain sensors, rain gauges, timers, and moisture sensors.

Chemigation-Irrigation

Chemigation is the injection of any chemical such as nitrogen, phosphorus, animal wastes, or pesticides into irrigation water and on to the land using the irrigation system (Anonymous 2012a, 2015b). Chemigation is recognized as a Best Management Practice (BMP). In order to use chemigation, the equipment, limitations and management must be understood. In North Dakota chemigation is almost exclusively limited to pivot sprinkler irrigation systems (Anonymous 2012a). Chemigation can also be used with other forms of sprinkler irrigation systems. Chemigation can save time, reduce labor requirements, and conserve energy and materials (Anonymous 2015a).

In Nebraska, the Nebraska Department of Agriculture (NDA) is responsible for certification and licensing of pesticide applications. The NDA does not regulate the chemigation process or facilitate where pesticides are mixed or loaded (Anonymous 2014). These two functions are regulated by the Nebraska Department of Environmental Quality (NDEQ). The Nebraska 23 Natural Resource Districts (NRDs) and the NDEQ regulate the Nebraska Chemigation Act that places certain requirements on anyone who uses chemigation in Nebraska. The NRDs inspect the safety equipment and chemigation systems and issue chemigation site permits. The NDEQ develops the statewide regulations and coordinates the overall program and issues applicator certification to trained individuals.

Chemigation in Kansas and other states requires compliance with federal and state pesticide laws and regulations, safety measures, knowledge of irrigation systems, chemigation, and equipment and calibration procedures (Anonymous 1985). Chemigation is used sometimes where application of a herbicide by ground or aerial equipment may cause drift drainage to nearby sensitive crops or plants. A pesticide applicator must demonstrate his/her competence by passing an examination. The applicator must demonstrate competence in pesticide application as well as chemigation procedures. Training programs are available in states where needed.

Water and Climate

Water limitations are one of the most important factors in producing crops and animal products even though an adequate soil media may exist. Climate determines the types of plants and animals that inhabit the region (Spencer 2003). Climate includes long-term averages of weather patterns that tend to repeat year after year. Climate is primarily considered as daily temperatures and types of precipitation that may occur. However, average hours of sunshine, wind direction and velocity, the number of frost-free days and measurement of extreme air temperature and fog-free days are also recorded. Weather forecasts and predictions are important to everyone, but especially to farmers since it influences the time of planting, crop and animal growth, harvest, and overall activities. Agriculturalists can review past weather records, but variation in air temperature and precipitation can be unpredictable for future use. Climate influences the quality and quantity of crops and animals produced in a given area and the size of the human population it can support.

Brady (1990) stated that about one fourth of the land area in the world supports enough vegetation to provide grazing for animals, but cannot be cultivated. But there is another 25% or one fourth of the land area of the world that can be cultivated.

Major areas of arable soil per person include North America, the Soviet Union, and Oceania and least in Asia, Europe and Africa.

Climate and water play a vital role in crop and animal production in these areas. Mineral deficiencies are severe in Southeast Asia and South America. Monsoon rains can also cause excess water in Southeast Asia. Shallow soils are problems in North, Central, and South Asia. Where water is available, arid lands can be irrigated.

World Irrigation

Irrigation covers about 200 million ha of agricultural land in the world (Anonymous 2012b). Asia has the largest acreage in the world. Egypt irrigates about 16,400,000 m^2. Vast acres are also irrigated in South Africa. In the U.S.A. only about 9% of agricultural land is irrigated but the U.S.A. ranks third in the world in regard to irrigated acreage. Mexico has about 16% of their land under irrigation. Russia has 10 million ha of irrigated land. Italy, Spain, and Southern France irrigate vast acreages around the Mediterranean region. South America, Brazil and Argentina irrigate a lot of agricultural lands as well as Peru. Irrigation in Australia is well-developed. Asia, China, India, and Pakistan account for most of the irrigated acreage.

Arable world lands (able to be plowed) in 2008 was 1386 million ha out of a total of 4883 million ha (Anonymous 2015e). Therefore, roughly 14.4% of the arable land of the world is under irrigation for cropland. This shows our dependence upon these lands to grow crops for human and livestock use and irrigation (water) becomes more important as human populations expand.

Water Use by Crops

A large part of the irrigation water applied to farms is lost by evaporation and transpiration (Blaney 1955). Consumptive use includes all transpiration from plants (discussed early in this chapter) plus evaporation losses from bare soil and water surfaces. Therefore, the amount of irrigation water needed for a specific crop includes the total water use, exclusive of precipitation. Rainfall is sometimes helpful during the growing season in reducing irrigation water requirements.

Temperature largely determines the types of crops that can be grown, and water use is determined probably more by temperature than any other factor. Evaporation and transpiration from land and plant surfaces are accelerated by low humidity and hot, dry winds passing over the crop area.

Latitude of the earth affects water use. Longer growing days may allow plant transpiration to continue for a longer period and produce an effect similar to that of lengthening the growing season in crop production. For example, many vegetables and flowers can be grown in Alaska during long days and a short growing season and efficient water use may be as productive as plants grown in the southern coast of California U.S.A. where there is more than 360 days of growing season and high water use.

Water use in the United States in 2010 for total irrigation withdrawals were 115,000 million (M) gallons/d which accounted for 38% of the total freshwater withdrawals. Withdrawals from surface water sources were 65,900 Mgal/d which accounts for 57% of the total irrigation withdrawals. Groundwater withdrawals for 2010 were 49,500 Mgal/d. About 62,400 thousand acres were irrigated in 2010 - 31,600 thousand acres (51%) with sprinkler systems, 26,200 thousand acres with surface flood, and 4,610 thousand acres with micro-irrigation systems. The national average application rate for 2010 was 2.07 acre-feet per acre.

The majority of total US irrigation withdrawals (63%) and irrigated acres (73%) were in 17 western states. Surface water was the primary source in the arid West. More groundwater was used in Kansas, Oklahoma, Nebraska, Texas, and South Dakota than surface water (USDI, USGS 2014).

Methods to Measure Irrigation Needs

A variety of methods and devices can be used to measure soil-water (Evans et al. 1996). They include the feel method, gravitational method, tensiometer, electrical resistance blocks, neutron probe, Phene cells, and time domain reflectometer.

The feel method involves estimating soil-water by feeling the soil, but is entirely subjective. This method requires a very experienced operator. The gravimetric method is determined by taking an appropriate soil sample and weighing it, drying it in an oven for 24 hours at 220°F, and then re-weighing it to calculate water loss. It is a good method to calibrate other measuring methods.

The tensiometer is an airtight water-filled tube with porous tip on one end and a vacuum gauge on the other end that indicates soil-water suction (negative pressure) usually expressed as tension on the vacuum gauge. Soil-water tension is usually expressed as bars or centibars. One bar is equal to 100 centibars (cb). Tensiometers work in the range of 0 to 0.8 bar and are affordable and easy to use. They work best in sandy and coarser-textured soils.

Electrical resistance blocks and meter work on the principle that water conducts electricity. When installed the water suction of the porous block is in equilibrium with the soil-water suction of the surrounding soil. As the soil moisture changes, water content of the block changes. The electrical resistance between the two electrodes in the block increases as water content of the block decreases. The block resistance can be related to soil-water content by a calibration curve. To make a reading, lead wires are connected to the meter containing a voltage source. The meter reads 0 to 100 or 0 to 200. High readings indicate high levels of soil-water and low readings indicate low levels. Resistance blocks are best suited for fine-textured soils such as silts and clays that retain at least 50% of their plant-available water at suction greater than 0.5 bars (50 cb).

Other methods include neutron probes which are easy to use, are reliable and accurate but are very expensive. A Phene cell is less expensive than a neutron probe. A neutron probe uses a radiation source to measure soil-water. A Phene cell works on the principle that a soil conducts heat in relation to its water content. By measuring the heat conducted from a heat source and calibrating the conductance versus water content for specific soil, the Phene cell can be used to reliably determine soil-water content. The time domain reflectometer (TDR) consists of two para rods or stiff wires inserted into the soil to the depth needed. The rods are connected to an instrument that sends an electromagnet pulse (or wave) of energy along the rods. Rate wave energy conducted reflected back to the soil surface is directly related to the average water content of the soil. The TDR is very expensive to use and is reliable. Technical assistance may be required for proper use of this equipment.

Knowledge of irrigation skills and chemigation is required both from a legal and practical point of view to ensure safety of the land, crop, and humans involved. Dedication and long work hours of the operator are required for success.

Preparing Land for Irrigation

On many farms some of the water may be wasted because fields are not properly prepared. Preparation for irrigation involves land leveling and reshaping the field surface to improve uniformity of water application. Surface leveling to provide surface drainage for even water penetration at irrigation or rainfall may be desired. Grading land to a uniform slope in the direction of irrigation and removing all slope at right angles to it should be the goal. Land graded to an exact level with no slope in any direction permits the most efficient irrigation, but it is not practical for all fields. It may also not be suitable in humid areas where excessive rainfall must be drained off the fields. Maintenance of leveled land is required to keep it leveled. Some land is too steep for irrigation and costs do not justify leveling.

In some fields leveling is done in strips or lands at different elevations separated by low ridges or borders (bench leveling). The type of leveling and layout of the field should always consider the crops to be grown and farming equipment to be used and irrigation method.

In planning a field to be leveled, a good design and engineering is required. Flood (furrow) irrigation uses gravity to transport water (USDI, USGS 2015). Farmers use leveling equipment, some guided by a laser beam, to level a field before planting. Since water flows downhill, water must flow evenly throughout the entire field. Water flow is controlled in some cases by "surge flooding" (water released at prearranged intervals to reduce unwanted runoff). A large amount of water may be wasted in runoff but can be captured in runoff ponds and reused by pumping it back to the front of the field for the next irrigation.

We discussed irrigation methods earlier in the chapter but "drip irrigation" is more efficient than flood irrigation because water is sent through pipes with holes in them. Evaporation is reduced compared to flood irrigation. Pipes can be laid in crop rows by their roots. Pipes can be above or below the soil surface. "Spray irrigation" is like watering the lawn at home. Large spray systems on farms are commonly center-pivot systems on metal frames on rolling wheels that hold the water tube out into the fields. There can be a very big water gun at the end of the tube. Electric motors drive the wheels that move each frame in a big circle around the field spraying water. They can be detected by the large green circles of the growing crop. Sprayed water is lost to the atmosphere in hot, dry weather and/or windy conditions in large amounts. Water gently sprayed from hanging pipes reduces water loss.

Crops Irrigated

Just like home gardens, just about any herbaceous plant or woody plant can be irrigated during the dry season or arid region for survival, beauty, or crop production. In the United States, Smith et al. (2014) lists barley, cotton, cool-season forages, lettuce and spinach, edible grain legumes, maize, peanut, potato, range and warm-season forage grasses, rice, sorghum, soybean, sugarbeet, sugarcane, sunflower, and wheat as major crops. They are major crops in other parts of the world also. With the exception of rice, lettuce and spinach, most can be grown in dry-land farming, but they can also be irrigated where irrigation can be provided.

Water Relation in Plants

As discussed earlier, the study of water in plants is a complicated science. Regions where rainfall is abundant and fairly well distributed over the growing season, abundant vegetation can be produced (Kramer 1983). Rain forests in the tropics and other rainfed areas attest to abundant plant growth. Where the growing seasons are dry and hot and desert conditions prevail, vegetation may be limited. Adequate water has physiological importance. Almost every plant process is affected directly or indirectly by the water supply. Fresh weight of herbaceous plants is 80 to 90% water and around 50% for woody plants during the growing season. Therefore, water is an important part of protein and lipid molecules that make up the protoplasm of the cells. Water also serves as a solvent in which gases, minerals, and other solutes enter plant cells and move from one cell to another and in one organ to another. Water is a reactant or substrate in many plant processes including photosynthesis. It also is essential in maintaining turgidity essential for cell enlargement and growth. Turgor is important for stomata opening and the movement of the leaves, flower petals, and various specialized plant structures.

There is strong interest in increasing drought tolerance, especially of some of our major crops. This can be done by plant breeding and/or genetically modifying the plant. Eventually plant breeders may produce characteristics that will increase tolerance of plant dehydration and water use efficiency without loss of yield by continuing research efforts.

Antitranspirants are applied films of material that cover the entire leaf surface and reduce water loss or application of substances that cause closing of stomata. However, by closing stomata and decreasing escape of water vapor, uptake of CO_2 by leaves is also decreased. Application of latex, polyvinyl waxes, polyethylene, and other chemicals has been tried with limited success in reduced water loss from plants or reducing water needs.

Drainage of Fields

American farmers have drained land since the early days of settlement (Wooten and Jones 1955). Early drainage works were established in swamps of Virginia, North Carolina, Delaware, Maryland, New Jersey, Massachusetts, South Carolina, and Georgia. Drainage work was carried out under the authority of colonial and state laws. Other states were involved later. Early work started in the colonial days with the use of tile systems and outlet ditches. Wooten and Jones (1955) indicated much land in the Ohio and Mississippi Valley could not be cultivated without drainage. Malaria was prevalent in large areas. Drainage added millions of acres in the United States that could not be farmed otherwise.

Under the Swamp Land Acts of 1849, 1850, and 1860 approximately 64 million acres of swamp and overflow land in 15 states were reclaimed for agriculture use (Wooten and Jones 1955). The Swamp Land Acts also included both flood control as well as drainage, but apparently the cost of the flood control and its disastrous effects led to huge increased expenditures under state law. Much of the Delta region was altered through drainage activities initiated and financed by creation of local drainage districts under state laws by 1900. By 1950 the agriculture census reported nearly 103 million acres of land in organized districts and county drainage enterprises in 40 states. This involved large expenditures for flood control and drainage improvements.

Today farmland is still being reclaimed by drainage, and tile systems are used to drain wet areas. Large dams provide irrigation water and help control floodwater. Open ditches, slopes, banks, dikes, levees, and canals bring or divert water to desired locations on or off farmland. Poor drainage and too much water can be as detrimental to crop growth as crops deficient in adequate water. Pumps are used to move water to desired locations. Floodwater cannot always be contained since excessive rainfall may overwhelm the system, but containing floodwater is in place in many communities. Waste and sewage water may also need attention and disposal.

The Importance of Groundwater

Thomas (1955) stated that the underground reservoirs of water contain the largest storage of freshwater in the nation - far more than all of the surface capacity including the Great Lakes. It has considerable importance indirectly because the minimum flow of the streams is sustained primarily by groundwater. The depletion of groundwater in some locations has resulted in depletion of streamflow and reduced water supply for non-withdrawal uses. Nebel (1987) predicted since irrigation is the greatest consumer of water, depletion of groundwater will have a great impact on our culture. This has already occurred since millions of acres in the Great Plains have been converted to dryland farming because of the depletion of the great Ogallala Aquifer. Depletion of the water table has dropped the water level to the point where irrigation has become increasingly costly as deeper wells must be drilled to compensate for the falling water table. As irrigation is abandoned, crop yields and land value are significantly reduced. Removal of groundwater may also allow the ground to settle resulting in severe property damage and loss of value (Nebel 1987).

Nebel (1987) estimates that in the United States agriculture consumes about 700 gallons per day per person, 600 gallons per day for irrigation and electrical power production, and another 520 gallons for industry and home use for flushing and washing. This amounts to a total of 1820 gallons of water per person each day in the United States. A person can survive on one quart of water per day. This does not account for water needed to maintain lakes, streams, and rivers. Therefore, availability of fresh water use for plants and animals, including humans, is very critical. Water management is needed to maintain groundwater and water supplies, which is difficult with heavy use. The other problem with groundwater is keeping it free of pollution and saltwater intrusion. The list of pollutants includes soil particles (sediment), abundant nutrients from fertilizers, pesticides, fecal wastes, oil and grease from vehicles or mining, human litter, sewage, toxic wastes (heavy metals), radioactive materials, or any compound or organism that may actively pollute surface and groundwater harmful to living organisms. Methods are available to prevent and/or remove toxic substances in water sources (Charbeneau et al. 1992).

Glennon (2002) indicated that water pumping disrupts the hydrological cycle using fresh water for billions of gallons of bottled water, irrigation, mining, fracking, and other human activities. This excessive pumping of our aquifers has created great environmental concern. It has caused rivers, springs, lakes, and wetlands to dry up. Glennon (2002) reported that in the Southwest rivers, such as the Santa Cruz in Tucson, dried up. Water pumping around Tampa Bay, Florida has turned lakes into mud flats and has cracked the foundation of homes. Glennon (2002) suggested that the Mecan River in Wisconsin and great trout fishing and local wells were at risk because water bottling companies and other uses can affect the water supply. A change in the water environment can affect wildlife (fish) and degrade the environment. Local opposition by residents has changed adverse water use to conservation in some areas.

Glennon (2002) recommends that states 1) craft water conservation programs; 2) establish minimum streamflow and protect flows from pumping of hydrologically-connected groundwater; 3) stop unregulated groundwater pumping; 4) tax water pumped from any well within a certain distance of a river, spring, or lake; 5) collect data on wells; 6) require a new pumper to mitigate impact on the environment; 7) use financial incentives (pay more as conditions dictate); 8) whenever a water rights transfer occurs, the state should require that a small percentage of the water be dedicated for environmental purposes.

Summary

Humans and their plants and animals depend on water for survival. In areas of adequate water, soil, and weather the human population has done well.

Plants need water for growth and transpiration (loss of water vapor). Transpiration keeps the plant turgid, cool, provides nutrients, and is part of the photosynthesis process.

The hydrological cycle consists of an alteration of evaporation and condensation. Once in the air, water molecules condense and eventually return to the earth as precipitation. Some is absorbed by soil, occurs as surface and runoff water, or becomes groundwater; but eventually it will re-evaporate and complete the cycle. Water is only 0 to 4% by volume in the atmosphere.

In the lower atmosphere nitrogen and oxygen content are most abundant. The upper atmosphere contains concentrations of electrons and ionized particles of oxygen and nitrogen. Humidity is the quantity of water held in a given volume of air.

The amount of precipitation that occurs in an area is the primary factor in determining the type of ecosystem. The distribution of precipitation over the earth is basically dependent upon patterns of heating and cooling of the earth's atmosphere.

Acid rain (carbonic acid, H_2CO_3) from carbon dioxide and water vapor results in rainwater of the pH of about 5.5. It results from volcanoes and burning of sulfur-contained coal. It can disrupt the well-being of plants and animals. The Clean Air Act in the United States helps control acid rain.

In areas deficient in rainfall, irrigation can be used to grow crops, gardens, and lawns. The five types of irrigation include 1) furrow, borderstrip, or basin irrigation; 2) drip, spray ,or micro-sprinkler or bubbler; 3) sprinkler or overhead; 4) sub-irrigation; and 5) manual irrigation. Chemigation is the injecting of any chemicals, such as fertilizers or pesticides, into the irrigation system and applied to the land.

A large part of the irrigation water applied to soil is lost by evaporation and transpiration. Therefore, the amount of water needed for a specific crop includes the total water used exclusive of rainfall. A variety of methods and devices can be used to measure soil-water when irrigation is needed. They include the feel method, tensiometers, electrical resistant blocks, neutron probes, Phene cells, and time domain reflectometers.

Land is prepared for irrigation by land leveling and reshaping fields and requires a good engineering design. Just about any herbaceous or woody plant can be irrigated during the dry season or arid region. Main crops include barley, cotton, cool season forages, lettuce and spinach, edible green legumes, maize, peanut, potato, rangeland and warm-season forage grasses, rice, sorghum, soybean, sugarbeet, sugarcane, sunflower, wheat, and others.

In the past 200 years in the United States, American farmers have drained land for farming on millions of acres. If too much water exists, plants do poorly and the water must be drained away from the land or abandoned for some other use. However, there is some concern about our underground reservoirs and fresh water supply. Groundwater is needed to sustain flow of streams. Depletion of groundwater has reduced streamflow and well water supply in some areas. Bottled water, irrigation, and other human uses are of great concern because it may destroy the desired uses of the hydrological cycle and can drastically change the environment. Steps must be taken to conserve, replenish, and avoid contamination of groundwater for the health and benefit of all humanity.

Chapter 3. Literature Cited

Frank B (1955) .The Story of Water as the Story of Men. Pages I-8 in Stefferud A ed. Water. The 1955 Yearbook of the Agriculture, U.S. Department of Agriculture. Superintendent of Documents. U.S. Government Printing Office. Washington D.C.

Nebel BJ (1987). *Environmental Science: The Way The World Works.* 2nd edition. Prentice-Hall, Inc. Englewood Cliffs, New Jersey. 671 p.

Kramer PJ (1983). Water Relations of Plants. Academic Press, New York. 489 p.

Robbins WW, Weier TE, Stocking CR (1957). *Botany: An Introduction to Plant Science.* John Wiley & Sons, Inc. New York. 578p.

Galston AW, Davies PJ, Satter RL (1980). *The Life of the Green Plant.* 3rd edition. Prentice-Hall, Inc. Englewood Cliffs, New Jersey. 464 p.

Spencer EW (2003). *Earth Science: Understanding Environmental Systems.* McGraw-Hill Companies, Inc, New York. 518 p.

Anonymous (2009). Types of Irrigation. Retrieved from http://nasatwiki.icrisat.org/index.php/Types of litigation. Accessed 18 august 2015.

Anonymous (2015a) Types of Sprinkler Systems. Retrieved from www.conserveh20.org/sprinkler-system-types. Regional Water Consortium Providers. Accessed 15 August 2012.

Anonymous (2012a). Chemigation. Retrieved from www.ag.ndsu. edu/irrigation/chemigation. North Dakota State University Extension Service. Fargo, North Dakota. Accessed 26 August 2015.

Anonymous (2015b). Chemigation Irrigation Systems. Retrieved from www.growsman.com/chemigation. Lindsay Corporation, Omaha, Nebraska. Accessed 26 August 2015.

Anonymous (2014). Nebraska Chemigation Regulations FAQ. Nebraska Department of Environmental Quality. Retrieved from www.dep.state.ne.us/pubica.nsf/7efidde82e69a6fe86258e... Accessed 27 August 2015.

Anonymous (1985). Chemigation in Kansas. Certification Chemigation Equipment Operator Examination Manual CSL-31A (09-07) Kansas Department of Agriculture, Pesticide and Fertilizer Program, Topeka, Kansas. Retrieved from agriculture.ks.gov/does/ default-source/re-chemigation/... Accessed 27 August 2015.

Anonymous (2012b). Irrigation Regions. Retrieved from www.travel-university.org//irrigation-regions.html. Accessed 27 August 2015.

Anonymous (2015c). Arable Land. Retrieved from en.wikipedia.org/ wiki/Arable_land. Accessed 27 August 2015.

Blaney HF (1955) .Climate as an Index on Irrigation Needs. Pages 341-345 in Stefferud A ed. Water. The 1955 Yearbook of the Agriculture, U.S. Department of Agriculture. Superintendent of Documents. U.S. Government Printing Office. Washington D.C.

U.S. Department of Interior/U.S. Geological Survey (2014). Water Use in the United States. Retrieved from http://water.usgs.gov/watuse/wuri.html. Accessed 28 August 2015.

Evans R, Cassel DK, Sneed RE (1996). Measuring Soil Water Use for Irrigation Scheduling: Monitoring Methods and Devices. North Carolina Extension Service. Publication No. AG 452-2. Raleigh, NC. Retrieved from www.fae.ncsu.edu/programs/extension/evas/ag452-2.html. Accessed 28 August 2015.

U.S. Department of Interior/U.S. Geological Survey (2015). Irrigation Techniques. Retrieved from http://water.usgs.gov/edu/irmethods.html. Accessed 28 August 2015.

Smith S, Diers B, Specht J, Carver B (eds) (2014). *Yield Gains in Major U.S. Field Crops*. CSSA Special Publication 33. American Society of Agronomy, Inc. Madison, WI. 487 p.

Kramer PJ (1983). *Water Relation of Plants*. Academic Press, New York. 489 p.

Wooten HH, Jones LA (1955). The History of Our Drainage Enterprises. Pages 478-491 in Stefferud A ed. Water. The 1955 Yearbook of the Agriculture, U.S. Department of Agriculture. Superintendent of Documents. U.S. Government Printing Office. Washington D.C.

Thomas HE (1955). Underground Sources of Our Water. Pages 62-78 in Stefferud A ed. Water. The 1955 Yearbook of the Agriculture, U.S. Department of Agriculture. Superintendent of Documents. U.S. Government Printing Office. Washington D.C.

Charbeneau RJ, Bedient PB, Loehr RC (1992). *Groundwater Remediation*. Water Quality Management Library - Volume 8. Technomic Publishing Company, Lancaster, PA. 188 p.

Glennon R (2002). Water Follies: Groundwater Pumping and the Fate of America's Fresh Water. Island Press, Washington DC. 314 p.

AGRONOMIC AND GM CROPS

Introduction

A variety of nutritious, attractive, and productive crops are grown in the United States for human and livestock food. Markle et al. (1998) lists 691 crops from corn and wheat to edible legumes that provide human and livestock needs. This is an unbelievable number since most of us probably could only think of a few crops that may affect our lives or may not even associate crops with food sources.

Markle et al. (1998) lists acreages in the United States planted and/or harvested in the mid-1990s of wheat (69 million acres [MA]), corn (60-70 MA), soybean (60 MA), cotton (13 MA), sorghum (8 MA), and barley (7 MA) as the top six crops grown. The EPA (Anonymous 2013) lists the same crops in 2013 as Markle et al. (1998), except they exclude barley and list rice and hay crops with corn (84 MA), soybeans (74 MA), hay (56 MA), wheat (46 MA), cotton (10 MA), sorghum (4 MA), and rice (3 MA). The net worth of these crops are $143 billion with about $153 billion worth of livestock each year (Anonymous 2013).

Corn (84 MA)

The United States is the largest producer of corn (32% of world crop), grown on 400,000 farms. The United States exports 20% of the crop (Anonymous 2013). Corn grown for grain accounts for about one-fourth of the harvested crop acres in the United States. About 80% of all corn grown in the U.S. is consumed by domestic and overseas livestock, poultry, and fish. Each American consumes about 25 pounds annually of corn products or about 12% of the U.S. corn crop. Corn is used in a variety of industrial uses including ethanol production, paints, candles, fireworks, drywall, sandpaper, dyes, crayons, shoe polish, antibiotics, and adhesives. Only about 2% (6 MA) of the total harvested plants are used for silage production (Anonymous 2013).

Soybean (74 MA)

Over 3 billion bushels of soybeans are harvested from 279,110 farms in the U.S. The United States accounts for 50% of the world production and 50% of the oilseed products. About 30 million tons are consumed by livestock as soybean meal.

Hay (56 MA)

The United States produces 119 million tons per year - alfalfa primarily. Most is used for domestic consumption. Alfalfa can be bailed or made into cubes or pellets.

Wheat (46 MA)

Wheat is grown on 160,810 farms in the United States each year and exceeds 227 billion bushels per year (one bushel equals a dry measure of the container holding 8 gallons).

The United States produces 10% of the world's crop. Seventy percent is used for food, 22% for livestock feed, and the remainder for wheat seed. The U.S. has about 25% of the world export market. A majority of U.S. wheat comes from the Great Plains.

Cotton (10 MA)

Cotton is grown on about 18,605 farms in 17 southern states, mostly in California, Texas, and the southeast states. The U.S. produces 30% of the world cotton, and the export value is $7 billion. From the 15 million bales produced or 7.3 billion pounds of cotton, 75% is used in apparel, 18% in home furnishings, and 7% in industrial products. Cottonseed is fed to dairy cattle and poultry and is used in food products such as margarine and salad dressing.

Grain Sorghum (4 MA)

Grain sorghum was grown on 26,242 farms in 2013 and is used in food products and livestock feed. It is also used in wallboard manufacturing and biodegradable packaging materials. Grain sorghum is water efficient and more drought resistant than corn. Most is grown in the Great Plains. The U.S. exports almost half of the production and has 70 to 80% of the world export market. It is also used to produce ethanol.

Rice (3 MA)

Rice is grown on about 6,000 U.S. farms in the Arkansas Grand Prairie, Mississippi Delta, Missouri, Louisiana, and Texas Gulf Coast. It is also grown in the Sacramento Valley of California. Three types of long, medium, short varieties exist. The U.S. produced about 6 million metric tons in 2015 (Anonymous 2015). China and India are the largest producers of 146 million metric and 104 million metric tons, respectively (Anonymous 2015). About 10% of the U.S. crop is exported with most used in U.S. food products and pet food.

History and Development

Corn

Corn or maize (*Zea mays* L.) was cultivated in the Americas at least 5600 years ago by Native Americans (Markle et al. 1998). The exact origin of corn is unknown but probably occurred in Central America and Mexico. Corn is a warm-weather annual, deep-rooted, and requires abundant soil moisture for best growth. It can grow nearly 20 feet tall from a single stalk depending on growing conditions. It has large, smooth leaves that can be more than 2 feet long and 2 inches wide at mid-point. Normally one to three or more ears develop and the flowers and later grain kernels are borne on a receptacle, termed ear that arises mid-point along the stem. The grain is enclosed in several layers of papery tissue, termed husks. Strands of silk, actually stigmas (female) of the flower, emerge from the terminals of the ears and husks at the same time pollen (male) from the terminal tassels is shed. Pollen (wind-blown) contacts the emerged silk or stigma to initiate fertilization of the egg. The pollen germinates and a pollen tube grows down through the silk to the egg cell of the female flower. The male gamete fuses with the egg and the corn seed or kernel develops from the fertilized egg.

Most varieties of corn produce seed 100 to 140 days after seeding. Corn varieties vary in kernel size and shape. The kernel or seed can vary from 1/8 inch long and nearly round to popcorn ½ inch long and flatter in shape. The kernel consists of an outer covering made up of two layers, 1) an outer pericarp and an inner testa, or true seed coat; 2) the endosperm making up two-thirds of the total volume, mostly starch; and 3) the embryo, the miniature plant structure that develops into a new plant after planting and germination and growth. Corn oil is obtained from the germ (embryo) of the kernel.

The three major types of corn grown in the U.S. are grain or field corn, mainly fed to livestock and some for silage, and sweet corn used mainly as human food and popcorn.

Ebeling (1979) reported before hybrid corn, the seed for next year's planting would be selected out in the cornfield, generally in early October. Choice ears, thick, even, and filled to the top were selected. After ears were dried, a kernel or two was selected from each ear and tested for swelling or sprouting ability when moistened. The best achievers were identified, and the most promising ears were stripped to supply seed corn for the next spring planting.

However, the principal basis for improvement of corn was hybridization plus improved cultural measures. Former breeding and selection of corn consisted of isolating superior strains obtained through open pollinations. Because the distribution of pollen was not controlled, a limit was reached where repeated crossings gave no improvement. When corn varieties were repeatedly inbred for many generations, uniform "lines" could be obtained. These lines were low in vigor and yield. When the inbreds were crossed, the resulting hybrids were superior to the original open-pollinated varieties. Eventually, resistance to disease and insect pests were bred into strains of corn by selection and hybridization.

Ebeling (1979) indicated that an enormous amount of work and determination was required of early- and present-day workers to develop superior hybrids. Many excellent hybrid strains have been developed by cooperative efforts of industry (seed companies), the U.S. Department of Agriculture, and many state university and experiment stations. Hybrids have resulted in greater uniformity of plants and ears, disease and insect resistance, adaption for mechanical harvesting, local soils and weather conditions, and resistance to lodging. Corn hybrids have also allowed improvements in percentage starch, protein, oil content and energy content, and yield. High-lysine content has been introduced into many well-established corn hybrids. Lysine is an essential amino acid required for optimum health and development in humans and most animals. High-lysine corn resulted from a mutant high-lysine gene called Opaque-2 (Ebeling 1979). High-lysine corn contains 69-100% more lysine and 60% more tryptophan than other hybrids (Ebeling 1979). A hybrid is an individual produced from genetically different parents (Bailey 2006). Natural or artificial processes can lead to the formation of a hybrid. The hybridization of normally self-pollinating species like corn involves the removal of the anthers (emasculation) and the artificial transfer of pollen from another plant (Bailey 2006).

Obviously the large volume of single-cross seed required each year cannot be obtained by hand-pollinated controlled crosses. The corn breeder chooses a well-isolated field and plants two rows of one inbred designed to bear the seed, alternated with one row of inbred male parent. Tassels are removed from those plants expected to bear hybrid seed, and the only pollen in the field is produced by the inbred male plants. The latter are also fertilized themselves and produce another generation of inbred seed. Two carefully selected and previously proved single-cross hybrids can be combined to produce a double-cross type; likewise a three-way hybrid can be produced. Strains of corn can be developed that are not only superior producers but are well adapted to particular conditions of weather, soil, disease, and insects - custom-built for particular needs. Some hybrids can produce much greater yields during short growing seasons than were possible with open-pollinated corn, and this has resulted in corn being successfully grown in areas farther north than was possible before the advent of the hybrid strains (Ebeling 1979).

The combine, as modified for harvesting corn, is important in U.S. agriculture in providing high efficiency in the corn harvest (Ebeling 1979). Corn ears mechanically harvested can be stored in an outdoor corn crib with wooden slats or mesh wire to keep them well ventilated. Others store shelled corn in an elevator like wheat or other grains. If corn is fed to livestock, it is generally harvested with a tractor-drawn chopper that picks up green cornstalks from windrows in the field and cuts the stalks and ears into small pieces. The chopped-up material is blown into a trailer attached to the chopper, then hauled to unload into feed bunkers or large silos. Some of the corn crops may be harvested for silage. The silo is a tall cylindrical structure that seals out air when filled with silage corn. The silo can store corn, grass, legumes, and other plants without spoilage. Silage can increase the nutritional value and palatability of the feed for livestock.

Smith et al. (2014) reported that maize hybrids released by DuPont Pioneer from 1930 to 2011 had a genetic gain of 1.4 bushels/acre/year (87.6 kg/ha/yr) planted at optimum density in well-watered environments and 0.8-0.9 bushels/acre/year (49.9-59.1 kg/ha/yr) when in drought-stressed environments. Genetic gain contribution, well-watered environments, and resistance to European corn borer helped gains in grain yield. More effective genetic diversity and crop management will allow U.S. maize breeders and farmers to make further corn yield gains and accommodate climate change for the foreseeable future, according to the authors.

We discussed very briefly the production of corn. It does not include food processing, marketing, management, ethanol production, or training of individuals in plant breeding, culture, and management or pest control. The purpose of this book is to show all the hard work and knowledge that is associated with each major crop we depend upon for life. It is essential that many qualified individuals be responsible for each crop to make certain the best seed or plants are available to farmers for the highest nutritional quality and yield, resistance to pathogens, insects, and weeds and are adapted to the soil environment in which it will be grown. Constant vigilance of these crops are essential for their maintenance by industry, the USDA, and state institutions through selection, plant breeding, and best management practices. Crop pests of pathogens, insects, nematodes, weeds, and other pests essentially have their own natural breeding programs that adapt to new hybrids and varieties of crops and attack the plants to reduce vigor, growth, and crop yield. When this occurs, new cultural methods or breeding programs are continually needed to keep these crops productive and safe.

Soybean

Markle et al. (1998) indicated that soybean has become the most important source of vegetable oil in the U.S. The plant is a bushy, hairy annual herb up to 3 feet tall. The hairy pods grow in small clusters with 2 to 4 seeds per pod. Seeds are generally about ¼ inch long. Pods are closed until the seeds are threshed out. They contain up to 25% of drying oil. Soybean oil is used as salad and cooking oil and in the manufacture of margarine and many other products such as soymilk, tofu, soy sauce, and salad oil.

Soybeans (*Glycine max*) originated from Asia (Ebeling 1979; Markle 1998; Specht et al. 2014) and was originally grown in the U.S. as a forage crop until 1941 when the area harvested for seed finally exceeded the area grown for all other purposes (Specht et al. 2014).

The area of soybeans harvested in 2012 exceeded 70 million acres (29 million hectares). The increase in U.S. total soybean production in the past three decades was coupled with a geographic shift northward and has become concentrated in the north central United States (Specht et al. 2014). In the United States the rate of on-farm yield improvement from 1924 to 2012 is 23.3 kg/ha/yr for soybean. Spect et al. (2014) indicated that comparison of on-farm improvement and genetic yield improvement in high-yield irrigated production environments suggests that about two-thirds of the on-farm improvement is due to release of higher yielding cultivars. Advances in better crop management also contributed to yield improvement in soybean.

Hay

The next highest number of acres grown in the United States after soybeans is hay at 55.7 million acres (Anonymous 2013). Hay includes primarily alfalfa but there are many acres of bermudagrass and many range and pasture forages that can also be made into hay. Alfalfa (*Medicago sativa L.*) originated in southeast Asia and was probably first cultivated in Iran (Heath et al. 1975). The first attempt to grow alfalfa in the U.S. was in Georgia in 1736 but it first succeeded when grown on the West Coast in California in 1850. It spread from there to other states. Alfalfa imported from Germany in 1857 paved the way for expansion of alfalfa in the north central states.

Alfalfa is grown worldwide because it is adapted to a wide range of climatic and soil conditions. Alfalfa is sometimes called the "Queen of the Forages" and is one of the most important forage plants in the U.S. (Heath et al. 1975). Alfalfa produces more protein per acre (ha) than any other crop for livestock and is high in mineral content and vitamins. It is used as hay, pellets, or low-moisture silage.

Brummer and Casler (2014) indicated that yield improvements of most forage crops have been limited. Data for alfalfa shows some gains in yield through breeding programs but not as effective for those in major grain and oilseed crops. Gains for alfalfa have been about 2 to 4% in yield per decade and is attributed to improvements in persistence and disease resistance.

Estimates of yield improvement with clover and grass breeding programs have also been limited using the criterion range of up to 6% per decade. Brummer and Casler (2014) indicated optimism in forage crop yield gains for future breeding programs.

Heath et al. (1975) reported that there are more than 80 cultivars of alfalfa produced in the U.S. The alfalfa seed production industry has become specialized and the latest cultivars are available to farmers. Bees trip the alfalfa flowers to produce seed. The tripping can be done by many insects; but insects like lygus bugs, the spotted alfalfa aphid, and alfalfa weevil are harmful to alfalfa.

Cool-Season Legumes and Grasses

Cool-season legumes include alfalfa, red clover, white clover and birdsfoot trefoil as well as numerous annual clovers and medics (Brummer and Casler 2014). Red clover is used in the U.S. for hay, pasture, green manure, silage, and as a cover crop (Markle et al. 1998). It originated in the Eastern Mediterranean Region but is now grown in Europe and the edge of the Great Plains east to the Atlantic and west to the Rocky Mountains in irrigated areas. White clover is similar to red clover in place of origin and uses.

Birdsfoot trefoil is native to Europe, North Africa, and Asia. It is used for pasture, hay, silage, soil improvement and road bank stabilization in the U.S. Birdsfoot trefoil is grown mainly in Oregon, but is promising for the southeastern states. All these clovers produce seed, are pollinated by bees, and are involved in breeding improvement programs.

Cool-season grasses include perennial ryegrass, Italian ryegrass, tall fescue, orchardgrass, timothy, smooth bromegrass, and reed canarygrass (Brummer and Casler 2014). Forage yield gains from 1950 to 2000 per decade showed small gains for the ryegrasses, tall fescue, orchardgrass, and timothy but none for smooth bromegrass and reed canarygrass. Italian (annual) and perennial ryegrasses were introduced from Europe and are used for hay and pasture in the Pacific Northwest. The ryegrasses can also be used as silage and a turf grass. Italian ryegrass is used as a cover crop on road banks, golf courses, and lawns. Tall fescue, also introduced from Europe, is used in the U.S. for hay, pasture, turf, and erosion control. A number of improved varieties are available. It is used in northern and southern states.

Orchardgrass, also from Europe, has been cultivated in the U.S. for over 200 years. It is a long-lived perennial bunch type and is found throughout the U.S. for pasture, hay, silage, cover crops, and ornamentals. Timothy is native to Europe and is a perennial bunchgrass important for hay. Over 70 cultivars are available (Markle et al. 1998). Smooth bromegrass is also native to Europe and Asia. The plant is a long-lived perennial with creeping rhizomes. It is used in erosion control, pasture, and hay and is widely grown in the U.S. Reed canarygrass is a perennial native to the U.S., Europe, and Asia. Its seeds are used for bird feed. It can grow on wet land and uplands and is widely grown in the North Central states, the Pacific Northwest, and Alaska. Improved varieties are available. It can be harvested for silage and hay.

Efforts to improve switchgrass as a forage crop have been in place since the 1950s (Brummer and Casler 2014). Breeding switchgrass for biomass yield became primary for the U.S. Department of Energy Bioenergy Feedstock Development Program (BFDP) (Brummer and Casler 2014). Use of improved variety selection, sustained selection and breeding, utilization of the entire growing season, genomic selection technologies, and production of hybrid varieties with heterosis (hybrid vigor) has been used for switchgrass improvement using classical and modern breeding strategies (Brummer and Caster 2014). Substantial biomass gains have been obtained. There are many other broadleaf plants and grasses that could be included as cool-season.

Rangeland and Warm-Season Forage Grasses

The development of rangeland grasses, namely the wheatgrasses and wildryes, and warm-season grasses was started in the late 1800s to the early 1900s (Jensen and Anderson 2014). Jensen and Anderson (2014) indicated that range grasses have unique breeding challenges and are very complex. Another factor confounding range grass breeding is the occurrence of significant genotype x environment interactions. U.S. rangelands have highly variable soils, precipitation, and temperatures making predictable annual yield difficult. Selection and release of lines selected from ecotypes have been done for some native species with limited success. Since the 1950s basic research on modes of reproduction and genetic relationships within and among different species and genera allowed breeders to explore the use of intraspecific and interspecific hybrids and interploidy breeding through chromosome doubling. New range grass breeding programs have resulted in increased seedling establishment, stand persistence of seed and forage yields, and improved forage digestibility. Species of concern include many cultivars of crested wheatgrass, intermediate wheatgrass, tall wheatgrass, bluebunch wheatgrass, Snake River wheatgrass, thickspike wheatgrass, slender wheatgrass, western wheatgrass, Russian wildrye, Basin wildrye, breadless wildrye, and altai wildrye. Many have originated from the former Soviet Union.

In the South basic breeding techniques and field trials have resulted in significant yield gains in bermudagrass over time.

Wheat

Wheat (*Triticum spp.*) is a cool-season crop that does best at preharvest temperatures averaging about 60°F with a minimum frost-free growing season of about 100 days (Markle et at. 1998). It is grown on about 46 million acres in the United States (Anonymous 2013) and on over 549 million acres worldwide. It is the most important food grain of the temperate zones both north and south. Carbonized grains have been found in Iraq and Syria dating back as early as 8000 B.C. The Middle East is probably the area of origin.

Wheat is an annual grass grown mainly as a winter annual in milder climates by seeding in the fall and harvesting from June through August. In rigorous winter climates it is mainly spring seeded.

The wheat grain or kernel is roughly egg shaped from 1/6 to 2/5 inch long and is made up of three parts. The outer covering is several cell layers thick and is the bran. It is separated during milling. It is about 12% of the kernel by weight. The endosperm is mainly starch and is 85-86% of the kernel. The germ or embryo expands into the new plant during germination and makes up about 2.5% of the kernel and is usually separated out during milling.

Wheats are classified as "hard" or "soft" depending upon the endosperm granularity. Wheat grown in the United States is classified into 10 species of *Triticum*. Six are cultivated and four are non-cultivated. Winter wheat includes hard red, soft red, and white. Spring wheat includes hard red, durum, and white.

Wheat is used mainly for food (70%) but substantial quantities are used as livestock feed (22%) (Markle et al. 1998). Some is cut for hay or used as pasture.

Graybosch et al. (2014) indicated yield of wheat (*Triticum spp.*), a vital world food supply, rose steadily in the 20th century due to improved agronomic practices and plant breeding. Averaged across all trials in USDA-ARS studies from 1980 through 2010, yield gains per year were approximately 0.8% and 1.2% from 1958 through 1980 per year. Future prospects for enhanced potential for grain yield may depend upon genetic engineering.

Cotton

Cotton (*Gossypium hirsutum L.)* is grown on about 9.5 million acres in the U.S. (Anonymous 2013) mostly in Texas, California, Mississippi, Arkansas, Louisiana, and Arizona. It is also grown in more than 30 countries and provides a major fiber source for textile manufacturers (Campbell et al. 2014).

Cotton is grown primarily for the fibers or lint, but the oil-containing seeds are highly important (Markle et al. 1998). Cotton, outside the tropics, is an herbaceous annual with fairly large, lobed leaves. The fruits are capsules which dehisced as they open. Each capsule contains up to 40 to 50 obovate, rounded or angular seeds to which the fibers or lint are attached. The lint and seeds are harvested from the dehisced bolls by hand or by a stripper or mechanical picker. The lint is removed from the seed mechanically at the cotton gin, then baled. Seeds are half hull and half kernel and contain 28 to 40% oil. A ton of seed yields about 300 pounds of oil. The meal or press cake is high-protein livestock feed. The oil for human use is used mainly for shortenings. It is also used for cooking and salad oils, margarines, and soap.

Cotton requires insect and weed control to obtain its yield potential. Without manual labor cotton production has been highly mechanized (Ebeling 1979).

Analyses of on-farm and replicated variety trials suggest that productivity and genetic gain of cotton yield has occurred since 1980 (Campbell et al. 2014). With control of the boll weevil and adoption of transgenic cultivars in 1996, cotton production has increased significantly.

Grain Sorghum

Sorghum (*Sorghum bicolor L.*) was grown on about 4 million acres in the U.S. in 2013 (Anonymous 2013). Its origins are from Africa and dates back to ancient Egyptian culture. Grain sorghum became more important once it was discovered that it could grow in drier areas of the U.S. (Markle et al. 1998).

Grain sorghum plants are coarse annual grasses. Nearly all that are grown in the U.S. are the dwarf types being under 5 feet tall (stems) and suitable for combine harvesting. The flowers and seed occur in relatively dense panicles from 3 to 20 inches in length and 3 inches wide at the apex of the main stem. The kernels are small and round to broad-conic in shape, similar to wheat in size. The grain is about 6% bran, 10% germ, and 84% endosperm, which is largely starch. Protein content is higher than corn but about equal to wheat.

Most sorghum consumed in the U.S. is used for livestock feed for cattle, swine, and poultry. Most exported is probably used as human food mainly to Japan, India and Europe.

A small portion of grain sorghum is made into starch for use in foods in the U.S. such as dextrose. Other uses include adhesives, sizing paper and fabrics and mud in oil drilling (Markle et al. 1998). Green grain sorghum plants contain prussic acid (HCN) which is poisonous to livestock and cannot be grazed. Several grain sorghum varieties are classified in seven agronomic groups.

Rice

Rice (*Oryza sativa L.*) is the principal food crop in practically all the tropical regions of the world and many subtropical areas (Markle et al. 1998). Although rice is grown on only about 2.6

million acres in the U.S., mainly in Texas, California, Arkansas, and Louisiana, it is a major crop in China (Anonymous 2013). Acres grown in China are about 100 million plus an additional 200 million acres in the world outside China. Therefore, about half the world depends upon rice for a major food source.

Rice is an annual crop. Within a few days under water and moist soil, plants germinate and emerge from the soil and branch from buds in the leaf axils near the soil surface. The number of tillers or branches and density of the plant stand varies with the variety used. Only tillers that develop early produce panicles and seed. Stem height ranges from about 1.5 to 6 feet or more. The panicle or seed head resembles oats but is more compact. The panicle is 4 to 10 inches long, branches and droops some. The seed contains a hull that is removed in milling and comprises about 20% of the kernel weight. Further milling removes the bran (5 to 8%) or seed coat, the germ (4%), and some of the endosperm (68%). This results in about an additional 10% removal by weight.

The primary use for rice is human food (Markle et al. 1998). The hull may be removed in oriental countries by use of a hollow block and wooden pounder which leave the bran and germ on the kernel. In the U.S., European countries, and also in the Orient, rice is machine-milled to remove the hull, bran, germ and part of the endosperm.

Milled rice is used mainly for direct consumption usually after cooking by boiling. Rice is also used in breakfast foods and in manufacture of alcoholic beverages. The bran is primarily livestock feed. Rice hulls are used for fuel, insulation, mulch and in certain manufacturing processes. Rice oil can be extracted from brown rice for salad dressing, cooking oil, and soaps. Glutinous rice, also called nomi, waxy or sweet rice, is used for desserts.

Seeding of rice to harvest takes about 110-130 days for early varieties and 195 days for late-maturing varieties.

McKenzee et al. (2014) indicated that in the U.S. from 1981 to 2011 average rice yields increased 86 kg/ha/yr to a 3-year national average of 7.8 tons per hectare. Future prospects for U.S. rice yield enhancement, including hybrids or transgenic rice for improved yield and grain quality, look promising. Regulatory restrictions on water, fertilizer, agrochemicals and the environment and detectable arsenic levels in rice are of concern.

Barley

Blake et al. (2014) indicated the barley (*Hordeum vulgare* L. spp. *vulgare*) in 2012 was harvested from 1.3 million ha (3.2 million acres) in the U.S. About 3.1 million ha (7.7 million acres) were grown in the U.S. from 1912 to 1927 and has steadily declined in the area planted due to the Conservation Reserve Program (CRP) and other reasons. This allowed marginal cropping fields in the Midwest and Upper Plain states to be rested. However, from 1942 to 2012, average yield gain was 43 kg/ha.

Blake et al. (2014) reported that the U.S. malting and brewing industries are the largest consumers of the U.S. barley grain crop and purchase more than one-half of the harvested crop. Blake et al. (2014) indicated that no-till cropping systems, improved soil nutrition and growing barley on better soils contributed significantly to barley yield gains over the past two decades in addition to improved genetic gains.

The other half of barley grown in the U.S. is used for livestock feed (Markle et al. 1998). Barley is nearly equal to kernel corn in nutritive value. It is especially valuable as hog feed. The entire kernel is fed after grinding or steam rolling.

Barley bran is used in cooking, cake, cookie, and bread.

Barley is one of the most ancient cultivated grains. Grains found in pits and the pyramids in Egypt indicate barley was cultivated there more than 5000 years ago. Barley is native to the Middle East and was first grown in the U.S. in 1602 (Markle et al. 1998).

Other U.S. Grown Crops

Other U.S. Grown Crops	U.S. Acreage (million)	Place of Origin
Peanut (*Arachis hypogaea* L.)	1.6	South America
Potato (*Solanum tuberosum* L.)	1.1	South America
Sugarbeet (*Beta vulgaris* L.)	1.2	Mediterranean
Sunflower (*Helianthus annuus* L.)	1.7	Western U.S.
Lettuce (*Lactuca sativa* L.)	<1	(Egypt) Middle East
Spinach (*Spinacia oleracea* L.)	<1	Central & Western Asia
Edible Grain Legumes	U.S. Acreage (million)	Place of Origin
Dry bean (*Phaseolus vulgaris* L.)	2	Central America
Dry pea (*Pisum sativum* L.)	<1	Turkey
Chickpea (*Cicer arietinum* L.)	<1	Syria and Iraq
Lentil (*Lens culinaris medikus*)	<1	Syria and Iraq

Many other field crops could be discussed but those included within are the major U.S. crops, most of which are grown worldwide. Cultural and genetic improvements and changes are constantly made to improve the quality and quantity of these crops in the U.S. and abroad (Thompson 2010; Edme et al. 2014; Hulke and Kleingartner 2014; Jansky et al. 2014; Holbrook et al. 2014; Panella et al. 2014; Simko et al. 2014; Vandermark et al. 2014).

Genetically-Modified Crops

Genetic engineering or genetic manipulation is the isolation of useful genes from a donor organism or tissue and their incorporation into an organism that does not normally possess them (Bailey 2006). The foreign DNA can be introduced into the host by means of a vector (carrier), usually a plasmid, bacteriophage, or other kind of virus. Genes for herbicide resistance enable crop plants to be treated with herbicides without injury and control associated weeds or render crops unpalatable or even fatal to insect pests. It is also used in medicine to produce antibiotics from microorganisms.

Genes contain pieces of DNA code which regulate all biological processes in living organisms (FAO 2004). The entire set of genetic information of an organism is present in every cell and is called the genome. Genetic engineering permits transfer of genes between organisms that are not normally able to cross-breed. For example, a gene from a bacteria can be inserted into a plant cell to provide resistance to insects in the plant. This is called genetically modified or transgenic.

Thompson (2010) poses the question of with more people on earth each day and all requiring food, shelter, fuel, medicine, and other life essentials, can they be sustained? This will require greater production from plants and animals with possible increased loss of land and water resources from increased human population.

For plants to adapt to changing conditions, they must inherit variation upon which selection can occur. The plants and animals we use today are the products of thousands of years of crossing and selection in different environments. However, today we can work with the DNA code from any organism and investigate new and hopefully improved organisms in the laboratory and field. Genetically-modified (GM) plants now occur in many U.S. crops such as herbicide-tolerant soybeans, corn, and cotton. Pest-resistant traits hopefully improve crop performance by protecting the plant from fungi and insects in plants like potato and tomato. Some GM tomato plants can be grown in seawater that produce very low salt concentrations in the fruit. GM crops can be produced to add nutrients and other desired qualities for human and livestock consumption.

Despite the limited numbers of genes introduced into commercial crops, many genes have been investigated in the laboratory and field. This is a difficult, time-consuming, and costly process. Like the pharmaceutical companies, agricultural companies must be able to recoup their cost and make a reasonable profit and protect their intellectual property. U.S. farmers have become highly dependent upon the seed companies and may not produce seed for personal use for many reasons.

Gene seed banks in many parts of the world may also hold plant materials suitable for plant variation and needs for the future. These banks also contain wild and/or weed species that may be useful and must be preserved even if they seem outdated.

Some negative elements exist with GM crops in that some people are concerned about the genetic makeup and negative nutritional effects of GM crops, although those supporting GM crops indicate they are safe. GM crops may cross-pollinate with wild relatives and cause herbicide or some other tolerance or change in native plants or weeds. Great care is required before GM plants are released. The U.S. government does not require GM crop produce to be labeled at the market.

Seed Production

Without seeds civilization would disappear. Without a means to grow crops and many forage plants, life could not be sustained. Seeds are produced in many forms but the structure develops from the fertilized ovule in flowering plants (Bailey 2006).

Seed usually contains one embryo together with the food supply which may be contained in a specially developed endosperm (storage tissue) or in the cotyledons (first leaf or leaves of the embryo in seed plants) of the embryo (Bailey 2006). The embryo and endosperm are encased by a protective coat, the testa. In gymnosperms the seeds remain naked and unprotected. To this class belong the firs, pines, spruces, and hemlocks (Robbins et al. 1957). In angiosperms the seed is enclosed within the ovary wall. The two groups of angiosperms are called monocotyledoneae (grasses) and dicotyledoneae (broadleaf plants). The crops we have discussed in this chapter are angiosperms. The reproductive process of corn seed has been given earlier in the chapter.

As explained earlier, the plant cell carries the chromosomes that determine inheritance. There are two crucial stages of chromosome behavior associated with sexual production: fertilization and meiosis. In fertilization two protoplasmic bodies, usually from different individual plants, unite to form one body of protoplasm, thus combining factors that determine characteristics from two different parents. In meiosis, four cells are formed, each containing sets of chromosomes that differ somewhat from each other in conventional reproduction (Robbins et al. 1957). This all takes place in the flower.

In flowering plants, seed or seeds develop as a result of fertilization in the flower. The seed or fruit can be used as food for human, wildlife, livestock, and many other organisms. Planted seed must germinate to produce new seedlings and adult plants for seed harvest, hay, or pasture. Good germination (high percentage) of seed is required for crop planting. Germination is initiated by imbibing water after planting in moist soil. Growth of the seedling plants to maturity also requires adequate water and nutrients. Germination of seed requires favorable moisture, temperature, and sometimes light resulting in release of food reserves for the developing plant. The radicle (root) of the embryo is usually first to emerge followed by the plumule (shoot). Those interested in the flowering parts of the plant required to make seeds are referred to Robbins et al. (1957) or any botany reference.

Importance of Bees and Other Insects

The maintenance of plant life generation after generation may be accomplished by asexual reproduction (the formation of buds, bulbs, and tubers) or by sexual reproduction (Metcalf and Metcalf 1951). In higher plants as discussed earlier in this chapter, the male gamete (pollen) unites with the female gamete (stigma) to form a new individual or seed through the process of pollination. Wind-pollinated flowers occur in corn, wheat and other cereals, nut trees, willows, oaks, pines, etc. Most of our fruits and ornamentals, however, such as beans, peas, tomatoes, melons, squash, clover, buckwheat, cotton, tobacco, and alfalfa depend upon insects to pollinate the stigma to initiate fertilization. Honeybees and bumblebees are important pollinizers.

Seed production of many of these crops would fail without bees. Many kinds of flies, butterflies, moths, and beetles also cross-pollinate many kinds of plants. Without bees plants like broccoli, asparagus, cantaloupes, cucumbers, pumpkins, blueberries, watermelons, almonds, apples, cranberries, and cherries would no longer be available. Honey would also disappear.

Names of Crops

Field, horticulture, forage crops, fruit, and nut trees all have names. A cultivar is defined as any variety or strain produced by horticulture or agriculture techniques not normally found in natural populations; a cultivated variety (Bailey 2006). Special or distinctive names are usually assigned to the crops. For example, head lettuce could be Great Lakes, Imperial, Iceberg, Crisphead, Woju, or cabbage lettuce. Club wheat could be variety Cluster, Dwarf, or Hedgehog (Markle et al. 1998).

Requirements for Crops

Field crops are usually tender and subject to damage and injury by storms, high wind velocity, hail damage, floods, drought, insects, weeds, birds, grazing animals, and other problems, especially when they are first getting established. As discussed in Chapters 2 and 3, good soil and adequate water are required for success in growing and harvesting field crops. Varieties and cultivars must be adapted and genetically suited for the environment in which they are grown. Adequate frost free days and favorable climate must fit the crop. In areas of good soil but limited water, irrigation may be necessary or desired. Crops tolerant of saline soils, drought, and excessive heat or cold are desired. For example, winter wheat planted in the fall must undergo cold winter temperatures (vernalization) before it will flower and produce seed the following spring and summer but can be damaged by extreme cold.

Once field crops are planted, weed control measures and possibly disease and/or insect control may be necessary to produce the crop. Crops such as cotton may require several applications of insecticides before harvest and chemical defoliation before harvest. Field crops need constant observation and care to produce during the growing season.

Crops are many times rotated between years to help control weeds, insects, or plant diseases or for other reasons. For example, corn may be followed by soybeans or grain sorghum because crop prices may favor the later crops. The land may need to be put back in pasture or hay crops for erosion control or changes in need for wildlife or livestock. Crops may be rotated to deal with agricultural chemical residues in the soil from one crop to another.

For grain crops adequate storage facilities are needed on the farm or commercial warehouse after harvest. Grain must be harvested and stored at low grain moisture (<12%) or it may rot and decay from microorganism activity. Facilities are available to dry grain if the standing crop must be harvested with high moisture content, but it is expensive. Vegetable crops such as lettuce and spinach are processed and used immediately after harvest unless they are grown for seed.

Some farmers grow various field crops for seed or planting use only. They take great care to produce high-quality seed that is high in percent germination and seedling vigor. Seed crops must be protected from bird, insect ,and disease problems and be free of weed seed. It must also be protected from frost damage. Farmers usually sell their certified seed to commercial facilities or resale the seed to other farmers and growers.

If seed or grain is not certified for planting, it can be used as human or livestock feed. Plant breeders work very hard to develop new hybrids and strains of forage crops, field crops, and floriculture and horticulture and tree crops that are resistant to pathogens and insects while attempting to improve their nutrient quality and yield. The scientists and agriculturalists go to great effort to ensure our food supply. Soil fertility, water management, and tillage practices were discussed in Chapters 2 and 3 for field crop production.

Summary

A variety of nutritious, attractive, and productive crops are grown in the United States for human and livestock food. The United States is the largest producer of corn at 32% of the world crop and it exports 20% of the crop grown each year. It was grown on 84 million acres in 2013. Soybean is next with 74 million acres followed by hay crops at 56 million acres. Large acreages of cotton (10 million), grain sorghum (4 million), and rice (3 million) are also grown.

Corn was grown in the Americas at least 5000 years ago by Native Americans. Today three major types are grown including field, sweet, and popcorn. The primary reason for improvement of corn yield was hybridization plus improved cultural measures. New varieties have improved the plant vigor and yield, resistance to disease and insect pests. Corn hybrids have also been improved in nutritional quality for humans and livestock, and corn grain has been involved in ethanol production.

Soybean is the most important source of vegetable oil in the U.S. Soybeans originated from Asia. The increase in U.S. production in the past three decades occurred as a result of geographical shift northward, irrigation, genetic improvement, and better crop management.

Hay crops are grown on huge U.S. acreages mainly due to alfalfa which originated in southeast Asia and was first believed to be cultivated in Iran. It was grown in the U.S. in the early 1700s. It is grown worldwide and is the "Queen of the Forages." It produces more protein per acre than most other crops used by livestock and is high in mineral content and vitamins. There are more than 80 cultivars of alfalfa, and bees are necessary to pollinate the flowers to produce seed.

Cool-season legumes include alfalfa, red clover, white clover, and birdsfoot trefoil as well as a number of other annual clovers and medics. All are important for hay, green manure, tillage, or as pasture for livestock.

Cool-season grasses include perennial ryegrass, Italian ryegrass, tall fescue, orchardgrass, timothy, smooth bromegrass, and reed canarygrass. They are important for hay, pasture, and silage.

The development of warm-season grasses mainly for rangeland are the wheatgrasses and wildryes. Selection and release of lines from ecotypes have been done for some native species with limited success. U.S. rangelands have highly variable soils, precipitation, and temperature making predictable annual forage yield difficult.

Wheat is a cool-season grass crop. It is grown on about 46 million acres in the U.S. and is the most important food grain of the temperate zones. Carbonized wheat grain has been found in Iraq and Syria dated back as early as 8000 B.C. Wheat is an annual grass grown mainly as a winter annual in milder climates and is classified as hard or soft depending upon the endosperm granularity. Wheat is a vital world food source.

Cotton is grown on about 9.5 million acres in the U.S., mostly in Texas, California, Mississippi, Arkansas, Louisiana, and Arizona. It is also grown in more than 30 countries. It is a major fiber source for textiles; but the oil-containing seeds are highly important for meal or press cake for livestock feed. The oil is also used for shortening, salad oils, margarines, and soap.

Grain sorghum originated from ancient Egypt, and most sorghum grain in the U.S. is used for livestock feed.

Rice is the principal food in tropical regions. The acreage is small in the U.S. (2.6 million acres), but in China about 100 million acres are grown. About half the world depends upon rice as a major food source.

Barley is grown on U.S. soil and the major use is for the malting and brewing industry and livestock feed. It has been found in pits and pyramids in Egypt and was cultivated there more than 5000 years ago.

Other U.S. grown crops discussed were peanut, potato, sugarbeet, sunflower, lettuce and spinach, and edible grains.

Genetically-modified crops, seed production, the importance of insects as pollinators, and requirements for crop growth were also discussed.

Chapter 4. Literature Cited

Markle GM, Baron JJ, Schneider BA (1998). *Food and Feed Crops of the United States.* 2nd Edition. Meister Publishing Co. Willoughby, Ohio. 517 p.

Smith S, Diers B, Specht J, Carver B, Editors (2014). *Yield Gains in Major U.S. Field Crops.* CSSA Special Publication 33, American Society of Agronomy, Inc. Madison, WI. 487 p.

Anonymous (2013). Major Crops Grown in the United States. Ag. 101. U.S. Environmental Protection Agency. Retrieved from www.epa.gov/Agriculture/ag101/cropmajor.html. Accessed 23 September 2015.

Anonymous (2015). World Rice Production 2015/2016. Retrieved from www.worldriceproduction.com. Accessed 28 September 2015.

Ebeling W (1979). *The Fruited Plain: The Story of American Agriculture.* University of California Press. Berkeley, CA. 433 p.

Smith S, Cooper M, Gogerty J, Löffler C, Borcherding D, Wright K (2014). Maize. Pages 125-171 in Smith S, Diers B, Specht J, Carver B, eds. Yield Gains in Major U.S. Field Crops. CSSA Special Publication 33, American Society of Agronomy, Inc. Madison, WI.

Bailey J (2006). *Collins Dictionary of Botany.* Harper Collins Publisher. Glasgow, Great Britain. 504 p.

Specht JE, Piers BW, Nelson Rl, Francisco J, de Toledo F, Fortion JA, Grassini P (2014). Soybean. Pages 311-355 in Smith S, Diers B, Specht J, Carver B, eds. Yield Gains in Major U.S. Field Crops. CSSA Special Publication 33, American Society of Agronomy, Inc. Madison, WI.

Heath ME, Metcalfe DS, Barnes RF (1975). *Forages: the Science of Grassland Agriculture.* 3rd Edition. The Iowa State University Press. Ames, Iowa. 755 p.

Brummer EC and Casler MD (2014). Cool-Season Forages. Pages 31-51 in Smith S, Diers B, Specht J, Carver B, eds. Yield Gains in Major U.S. Field Crops. CSSA Special Publication 33, American Society of Agronomy, Inc. Madison, WI.

Jensen KB and Anderson WF (2014). Rangeland and Warm-Season Forage Grasses. Pages 219-266 in Smith S, Diers B, Specht J, Carver B, eds. Yield Gains in Major U.S. Field Crops. CSSA Special Publication 33, American Society of Agronomy, Inc. Madison, WI.

Graybosch R, Bockelman HE, Garland-Campbell KA, Garvin DF, Regassa T (2014). Wheat. Pages 459-487 in Smith S, Diers B, Specht J, Carver B, eds. Yield Gains in Major U.S. Field Crops. CSSA Special Publication 33, American Society of Agronomy, Inc. Madison, WI.

Campbell BT, Boykin D, Abdo Z, Meredith Jr. WR (2014). Cotton. Pages 13-32 in Smith S, Diers B, Specht J, Carver B, eds. Yield Gains in Major U.S. Field Crops. CSSA Special Publication 33, American Society of Agronomy, Inc. Madison, WI.

Monk R, Franks C, Dahberg (2014). Sorghum. Pages 293-310 in Smith S, Diers B, Specht J, Carver B, eds. Yield Gains in Major U.S. Field Crops. CSSA Special Publication 33, American Society of Agronomy, Inc. Madison, WI.

McKenzie KS, Sha X, Moldenhauer KAK, Linscombe SD, Lyman NB, Nalley LL (2014). Rice. Pages 267-292 in Smith S, Diers B, Specht J, Carver B, eds. Yield Gains in Major U.S. Field Crops. CSSA Special Publication 33, American Society of Agronomy, Inc. Madison, WI.

Blake T, Blake V, Wiersma J (2014). Barley. Pages 1-11 in Smith S, Diers B, Specht J, Carver B, eds. Yield Gains in Major U.S. Field Crops. CSSA Special Publication 33, American Society of Agronomy, Inc. Madison, WI.

Thompson P (2010). *Seeds, Sex and Civilization: How the Hidden Life of Plants has Shaped the World*. Thames and Hudson, Ltd, London, 272 p.

Simko I, Hayes R, Mou B, McCreight JD (2014). Lettuce and Spinach. Pages 53-85 in Smith S, Diers B, Specht J, Carver B, eds. Yield Gains in Major U.S. Field Crops. CSSA Special Publication 33, American Society of Agronomy, Inc. Madison, WI.

Vandemark GJ, Brick MA, Osorno JM, Kelly JD, Urrea CA (2014). Edible Grain Legumes. Pages 87-123 in Smith S, Diers B, Specht J, Carver B, eds. Yield Gains in Major U.S. Field Crops. CSSA Special Publication 33, American Society of Agronomy Inc. Madison, WI.

Holbrook CC, Brenneman TB, Stalker HT, Johnson III WC, Ozias-Akins P, Chu Y, Vellidis G, McClusky D (2014). Peanut. Pages 173-194 in Smith S, Diers B, Specht J, Carver B, eds. Yield Gains in Major U.S. Field Crops. CSSA Special Publication 33, American Society of Agronomy, Inc. Madison, WI.

Jansky SH, Spooner DM, Bethke PC (2014). Potato. Pages 195-217 in Smith S, Diers B, Specht J, Carver B, eds. Yield Gains in Major U.S. Field Crops. CSSA Special Publication 33, American Society of Agronomy, Inc. Madison, WI.

Panella L, Kaffka SR, Lewellen RT, McGrath JM, Metzger MS, Strausbaugh CA (2014). Sugarbeet. Pages 357-395 in Smith S, Diers B, Specht J, Carver B, eds. Yield Gains in Major U.S. Field Crops. CSSA Special Publication 33, American Society of Agronomy, Inc. Madison, WI.

Edmé SJ, Suman A, Kimbeng C (2014). Sugarcane. Pages 397-431 in Smith S, Diers B, Specht J, Carver B, eds. Yield Gains in Major U.S. Field Crops. CSSA Special Publication 33, American Society of Agronomy, Inc. Madison, WI.

Hulke BS, Kleingartner LW (2014). Sunflower. Pages 433-457 in Smith S, Diers B, Specht J, Carver B, eds. Yield Gains in Major U.S. Field Crops. CSSA Special Publication 33, American Society of Agronomy, Inc. Madison, WI.

Food and Agriculture Organization (FAO) (2004). The State of Food and Agriculture Organization 2003-2004. Genetically Modified Crops. Green Facts. Retrieved from www.greenfacts.org/en/gma/index.htm. Accessed 28 October 2015

Robbins WW, Weier TE, Stocking CR (1957). *Botany: An Introduction to Plant Science.* John Wiley and Sons, Inc. New York. 578 p.

Metcalf CL, and Metcalf RL (1957). *Destructive and Useful Insects.* McGraw-Hill Book Company, Inc. New York. 1071 p.

ALTERNATE CROPPING AND ENERGY

Introduction

In the 1930's crop yields in the United States and other parts of the world were essentially the same. Since that time scientists and federal policies have helped U.S. farmers dramatically increase yield of major crops and other commodities. The farms are much larger and fewer farmers are needed in food production.

In the 1980's problems of pesticides and fertilizers detected in groundwater, soil erosion, pest resistance to pesticides, cost of production of specific crops, and the fact that some other countries have closed the productivity gap and are more competitive in international markets have made alternate practices in agriculture a consideration (National Research Council 1989).

Alternative agriculture examples include farming practices that attempt to reduce or eliminate chemicals, pesticides, and antibiotics for pest and disease control. Use of integrated pest management (IPM) (may include several practices) to reduce pest damage, improve crop health, decrease soil erosion, and add organic matter and soil nitrogen by planting legumes were considered. Genetic changes in crops to resist insect and disease problems and improve nutrient use were necessary. Crop rotations were also considered.

Any alternative practices adopted must be stable and resilient, reduce financial risk, and provide a buffer against drought, pests, or other natural factors limiting production.

Farmers can reduce pesticide use on cash grain crops by crop rotations that disrupt the reproductive cycle, habitat and food supply of insects and disease, and by altering timing and fertilizer placement or growing leguminous forages and cover crops in rotation with corn, soybeans, and small grains. Fruit and vegetable growers can dramatically reduce pesticide use by IPM programs. Subtherapeutic use of antibiotics in livestock can be reduced or eliminated without sacrificing profit when not reliant on extreme confinement rearing. Manure can sometimes be substituted for commercial fertilizers in crop production when available.

Plants for Energy

Because of genetic and cultural improvements in crop yield and production, crops such as corn, grain sorghum, and soybeans are being grown to produce alcohol as well as food and feed crops (Smith et al. 2014). Roundup ready corn, sorghum, and soybeans for weed control and crop price increases have greatly expanded acreage since the 1980s and 1990s in these traditional crops.

Cultivation of corn and soybeans reached record high levels in the United States following the biofuels boom of the 2000s. Lack et al. (2015) found that crop expansion resulted in substantial conversion of long-term unimproved grasslands and land that had not been previously used for agriculture into cropland for alcohol production. Corn was the most common crop planted directly on new land. The conversion to corn and soybeans could produce as much CO_2 (carbon dioxide) into the atmosphere as 34 coal-fired power plants or 28 million automobiles operating for one year. Modifications of the 2014 U.S. Farm Bill Sodsaver Provision and improving enforcement of the U.S. Renewable Fuels Standard need consideration by policymakers.

Holechek and Sawolhah (2014) recently reviewed the history and various energy sources as they are now used. Oil is the most common and superior energy source. The authors state that the term energy return on investment (EROI) is commonly used to compare energy sources. The EROI is the ratio of the amount of usable energy acquired from a particular energy source to the amount of energy expended to obtain that energy source. If the energy source has an EROI of one or less, it is not economically viable since there is no net energy gain from its extraction. Fossil fuels such as oil, coal, and natural gas are economical energy sources; but unconventional oil sources such as shale oil, shale gas, tar sands, and deep-water oil wells increasingly replace conventional oil and gas but with increased cost of extraction.

Although much progress has been made in renewable energy sources (biomass, solar, and wind), costs compared to fossil fuels are also high. Nuclear power is great, but cost and safety concerns limit its expansion. Rangelands provide water, livestock products, wildlife, and recreation to millions of people throughout the world as well as fossil fuels. What the contribution of rangelands will provide in biomass, solar, wind, and thermal power is yet to be determined.

In the last 100 to 200 years, mesquite and associated woody plants have become a woody weed problem on millions of acres in the southwestern U.S. Early native people depended heavily on mesquite for food, medicine, cosmetics, recreation, paints, dyes, cordage, cradles, weapons, shovels, furniture, construction materials, and fuel. In modern times mesquite provides livestock feed from pods produced, wood products, and fuel. Mesquite wood has been developed for livestock feed, an energy source for generating electricity, and possible ethanol production with limited success. Charcoal barbecue chips are being produced. Mesquite is also extremely important for arid and semiarid regions and on saline soils where it prevents desertification and soil erosion and improves soil stability and fertility. However, mesquite can be a huge weed problem for the ranching industry, and management and control measures sometimes are needed and have been developed (Bovey 2016).

In California, the forestry sector biomass energy producers have the potential to work together to reduce the state's net greenhouse gas emissions by generating renewable electricity (Trittmonn 2015). Under California's Renewables Portfolio Standard, one-third of the electricity provided by the state's utilities must be from renewable sources by 2020. To meet these goals utilities must depend on power producers such as solar, wind, and biomass. California forests will play a vital role in meeting the state's goals. Almost the entire publication of Volume 69 Number 3 - 2015 of California Agriculture, University of California (californiaagriculture.ucanr.edu) is devoted to growing, harvesting, and transporting forest biomass for energy production.

Perlack et al. (2005) evaluated the land resources of the United States to determine if it could produce a sustainable supply of biomass sufficient to displace 30% or more of the country's present petroleum consumption. This would require about 1 billion dry tons of biomass feedstock per year. The short answer was yes. About 36 million tons could be produced from forestlands and about 998 million dry tons from agricultural lands for a total of 1.3 billion dry tons. Biomass residues from forestland would come from logging, fuel treatment thinning and fuelwood, forest products wastes, urban wood residues, and forest growth. Biomass residues from agriculture would be from corn and small grains, soybeans, hay, silage, cotton, rice, fallow land, CRP land, and pastures. The authors indicated the goal of producing 1 billion dry tons of biomass could be met with modest changes in land use and still meet food, feed, and export demands of the United States.

Over the past 10 to 15 years many ideas and programs have been proposed for alternate energy to petroleum using sewage sludge, livestock dung, seed and pulp oils from sea buckthorn berries, seed oil from mustard, rapeseed, canola, cranberry, soybean and waste cooking oil, microorganisms, wheat and barley straw, corn stover, switchgrass and various woods and plants for biofuel production.

In 2006 more than half of the U.S. electricity was generated from coal. Use of coal today is being discouraged because of alleged pollution and health problems.

Many books were written on the end of the "so-called" oil age and future energy needs and possible alternative solutions in the early and mid 2000s. During this time period, rising cost of energy was the most frequent concern of global business leaders, although still of great concern 10 to 15 years later in 2016. Oil and gas production from fracking has temporarily and significantly reduced petroleum cost and significantly increased availability. Coal is less available, but cost of alternative energy sources compared to oil and coal is still very costly.

Other plants proposed for alcohol production include jojoba (*Simmondsia chinensis*), Eucalyptus (*E. botrojoides, E. ovata* and *E. globulus*), Mesquite (*Prosopis* spp.), Napier (Elephant) grass (*Pennisetum purpureum*), Switchgrass (*Panicum virgatum*), sugarcane (*Saccharum* spp.), Guinea grass (*Panicum maximum*), Gamba grass (*Andropogon gayanus*), Alemangrass (*Echinochloa polystachya*), Willow (*Salix* spp.), Cassava (*Mani-*

hot spp.), Milkweed (*Asclepia* spp.), Loblolly pine (*Pinus taeda*), sweetgum (*Liquidambar styraciflua*), cottonwood (*Populus* spp.), Sorghum (*Sorghum* spp.) and a host of other herbaceous and woody plants. Time will tell if biofuels will even become a major energy source.

Alcohol Production

Nebel (1981) states that yeast cells feeding on sugars and/or high starch grains under anaerobic conditions produce alcohol as a metabolic waste product by fermentation. The alcohol can be further concentrated by distillation by boiling and recondensing the alcohol.

Ethanol is most often converted from corn or sugarcane by utilizing enzymes to convert starches to simple sugars and yeast to ferment the sugars into ethanol. It can also be produced from cellulose by a hydrolysis process to break down hemicellulose and lignin (Anonymous 2015). Alternatively, a gasification process, heat, and chemicals can convert cellulose into syngas (carbon monoxide and hydrogen) which is reassembled into such products as ethanol, methanol, and higher alcohols.

Alcohol Processing Plants

A corn ethanol plant includes the following: 1) stored milled corn, 2) mash processing and saccharification, 3) fermentation, 4) by-product processing and 5) process and plant support systems.

A hydrolysis cellulosic ethanol production plant includes all of the corn ethanol plant components plus a high temperature, high pressure dilute acid to break down the hemicellulose and dissolve the lignin. The addition of enzymes can increase the yield of sugars.

A gasification cellulosic ethanol production plant includes: 1) fuel receiving, chipper and storage, 2) syngas production and cleanup, 3) catalytic syngas conversion, 4) distillation and puri-fication and 5) process and plant support systems (Anonymous 2015).

Ethanol facilities in the U.S. as of November 2015 had the capacity to produce 15.5 billion gallons of ethanol per year. Nebraska produces 1,780 million gallons or 12 percent of the nation's production. Other states producing ethanol include Arizona, California, Colorado, Florida, Georgia, Idaho, Illinois, Indiana, Iowa, Kansas, Kentucky, Louisiana, Massachusetts, Michigan, Minnesota, Mississippi, Missouri, New Mexico, New York, North Carolina, North Dakota, Ohio, Oregon, Pennsylvania, South Carolina, South Dakota, Tennessee, Texas, Virginia, Wisconsin, and Wyoming for a total of 32 states with Nebraska (Anonymous 2015).

The chemical formula for ethanol is CH_3CO_2OH. Blends of up to 85% ethanol with gasoline are considered alternative fuels under the Energy Policy Act of 1992. Most major auto manufacturers indicate that most vehicles can use E85 fuels.

The downside of ethanol production is that much farmland is needed to grow the plant materials and that producing biofuels requires more energy than the fuel can generate (Anonymous 2015).

Multiple Cropping

Multiple cropping is growing and harvesting more than one crop on the same piece of land in a year. Multiple cropping should increase food, feed, or biomass productivity. Apparently multiple cropping has been practiced since antiquity so it is not a new concept, but only recently has received scientific attention. Developing new and improved technology will depend on the experience and ideas of interested growers and scientists to achieve increased agricultural productivity while maintaining the land and water resources.

Multiple cropping consists of essentially two underlying principles - that of growing crops simultaneously in mixtures (intercropping) or of growing individual crops in sequence (sequential cropping) (Andrews and Kassan 1976).

United States

Eastern United States

In the southeastern U.S. double cropping has been possible by herbicide use, shorter season cultivars, favorable prices and no-tillage planting techniques (Lewis and Phillips 1976). Double cropping soybeans after wheat or barley is the most successful multiple cropping system for grain production in the southeastern states and southern corn belt. A forage producer may grow one or two silage crops in a small grain-corn or sorghum system.

Western United States

In the western U.S. the only areas normally suited for double cropping with present agronomic crops and varieties are those having growing seasons longer than 200 days with adequate irrigation water (Gomm et al. 1976). Native grasslands and mixed timber-grass ranges are important intercropping systems providing forage, meat, timber, and water. In Wyoming on irrigated farms, intercropping with a cereal grain and alfalfa as companion crop is a common practice. The cereal grain can be harvested early and legume or grass can be seeded into the cereal grain stubble. In north central Nebraska, rye (*Secale cereale* L.) can be seeded in some irrigated sandy cornfields by airplane in the standing corn and fall pasture. The rye plants help control wind erosion. Alfalfa can be seeded in irrigated corn to avoid a season's delay in its establishment.

In Kansas soybeans can be aerially seeded into wheat crops about the first of May if the soil is kept moist until the soybeans are well established. Multiple cropping is also widely used in Oklahoma in bermudagrass sod. In the Texas highlands, cotton and sorghum have been grown in alternating four row blocks perpendicular to prevailing wind to provide protection of the young cotton plants from wind damage. The system is also used to protect young peanut (*Arachis hypogaea L.*) plants in the Cross Timbers and Rio Grande Plains. Other examples of intercropping and sequential cropping are available for the western U.S. but little double or sequential cropping is done north of latitude 37°N or above 600 m elevation. The shorter growing season restricts the planting and maturing of soybeans or sorghum after small grain harvest and rainfall is usually inadequate to start and sustain a second crop (Gomm et al.1976).

Other Countries

Tropical America

Where sequential cropping is practiced, usually two or three crops are included in tropical America using short season crops such as vegetables. Also some grain crops, such as field bean and maize, are often harvested in the green stage to make way for subsequent cropping (Pinchinat et al. 1976).

In the intercropping system the central or primary crop can be fairly typical of a region. In Central America it may be maize, whereas in the Amazon Basin and lowlands of Colombia and Venezuela it may be cassava (Pinchinat et al. 1976).

Generally industrial large market oriented crops such as coffee (*Coffee* sp.), bananas (*Musa paradisiaca sapientum*), cocoa

(*Theobroma cacao*), sugarcane (*Saccharum* sp.), cotton (*Gossypium* sp.) and rice (*Oryza sativa*) are produced on large to medium farms. Basic food crops such as maize (*Zea mays*), bean (*Phaseolus* sp.), cassava (*Manihot esculenta*), quinoa (*Chenopodium quinoa*), and others are produced on small farms (Pinchinat et al. 1976).

Tropical Asia

Tropical Asia includes the area between the Himalayan Mountains and the equator from Pakistan and India to Indonesia, New Guinea, the Philippines, southern Taiwan, southeastern China, and mainland southeast Asia (Harwood and Price 1967). Dominant crops are determined by water supply. In dryer regions of India and Pakistan corn. sorghum, millet, and wheat are grown. Rice is grown in wetter areas of the same countries, and nearly all of east Asia with high rainfall grows rice.

There is a prevalence of small farms in tropical Asia. The average size is 1.8 hectares. On vegetable farms, intensity of vegetables produced increased as farm size increased (Harwood and Price 1976). Because tropical climate conditions are suitable for field crop production most of the year, cropping pattern alternatives are numerous but factors such as markets, labor, capital, rainfall, irrigation, soil, elevation, and temperature affect production and the type of crop grown. At higher elevation mostly vegetable crops are grown.

Larger farms (greater than 5 hectares) are more mechanized and may include sugarcane, oil palm (*Elaeis guineensis*), rubber (*Hevea brasiliensis*), coconut tea (*Camellia sinensis*), etc., timber, pineapple (*Ananas comosus*), corn (southeast Asia), wheat (India, Pakistan), and rice. Some farms have animals and crop production interactions for milk and meat, recreation or work.

Sequences for crops grown in pure stands are by far the most common form of multiple cropping. For paddy rice-growing, a single crop of rice with 5 to 6 months of high rainfall can be grown followed by a second crop of transplanted rice followed by a low-tillage broadcast mungbean or cowpea (*Vigna* sp.) (Harwood and Price 1976). For lowland rice areas, two rice crops are possible.

Intercropping of coconut, oil palm, and young rubber trees in the early years of plantation with annual crops can be done. Cassava, soybeans, peanuts (*Archis hypogaea* L.), corn, and other crops are used. Many intercropping and sequential cropping schemes are practiced, and relevant research for the Asian farmer is needed to evaluate and develop the best cropping systems.

Tropical Africa

Tropical Africa lies south of the Sahara Desert bounded by an imaginary line both north and south of the equator (Okigbo and Greinland 1976). The area extends over the entire equatorial belt of Africa and latitudinally to about 20°N and 26°S. The land area is about twice that of the U.S. and is diverse in relief, climate, vegetation, and crops grown.

Africa is also racially diverse with many linguistic groups (Okigbo and Greinland 1976). There are 40 countries represented by different cultural, economic, and political experience so there is great diversity in the number of crops grown and variations in cropping and farming systems.

With the exception of some parts of Africa where farming methods are of Europe and plantation agriculture. Multiple cropping is a major component of existing farm systems. Farming systems have included nomadic herding, shifting cultivation, rudimentary sedentary cultivation, special horticulture, and plantation agriculture.

Modern farming systems include mixed farming, livestock ranching, large scale farms and plantations based on natural rainfall and/or irrigation, large scale tree crops, and specialized horticulture of market gardening, truck gardening, and fruit plantations.

Research and observations on mixed cropping, patch intercropping, and relay cropping systems in Africa indicate that:

1. Traditional farming practices intercropping because it gives higher yield and greater returns than the same crop in pure culture.

2. Crop pests are minimized by intercropping versus sole cropping.

3. Investigations of intercropping should include: a) plant populations, b) length of growing season, c) plantation structure, d) length of life cycle, e) nutrient requirements, f) planting patterns and g) soil fertility.

4. Additional needed information on intercropping: a) erosion control, b) insurance protection, c) labor and harvesting costs, d) best location for adapted crop and e) mechanization (Okigbo and Greinland 1976).

Middle East

Double cropping has a long history in the Middle East associated with irrigation about 4000 to 6000 years ago in Babylon (Nasr 1976). It has been practiced in Egypt and other Middle Eastern countries where adequate water is available.

The important crops in Egypt are divided into three groups according to the cropping season. The agriculture year starts November 1 and ends the following October 31. In winter, winter cereals wheat and barley (*Hordeum vulgare*), legumes, winter onions, flax, and winter vegetables are usually planted from October to December and harvested from April to June. In summer, cotton (*Gossypium* sp.), summer cereals (rice, corn, and sorghum), summer onions, peanuts, sesame (*Sesamum indicum*), kenaf (*Hibiscus cannabinus*), and summer vegetables are planted from March to June and harvested from August to November. When the Nile floods in summer, corn, sorghum, rice, and summer vegetables are planted in July or August and harvested in October and November.

One of the most common rotations is cotton once every three years and a double cropping pattern every year, which could be cotton followed by a cereal or vegetable for two years and cotton again the third year.

Multiple cropping is practiced to some extent in the Middle East but most extensively in Egypt. The potential for more multiple cropping increases as more irrigation projects are developed. More research and good extension (education) programs are needed to educate farmers to practice multiple cropping (Nasr 1976). Other countries involved in multiple cropping include Iran, Iraq, Saudi Arabia, Syria, and Jordan in addition to Egypt.

Future Needs in Multiple Cropping

Future needs are to evaluate the best possible crops, cropping patterns, and sequences for a given area and region. This will require an adequate time period to evaluate crop rotations and the most favorable intercropping systems that are adapted to that region. Improvements in crop varieties and resistance to crop pests (weeds, insects, diseases, nematodes) would be desired.

Crops efficient in water use, fertilizer, and nutrient requirements and heat and cold tolerant would be helpful. Cropping patterns that disrupt pest life cycles and produce greatest potential yield is the goal. Maintaining beneficial insects and other beneficial organisms with minimal herbicide, insecticide, and other pesticide use reduces cost and environmental damage.

Cover Crops

This chapter would not be complete without a short discussion of cover crops. Cover crops have been used for centuries but have become more in focus because of their economic benefits.

Blanco (2023) indicated that there are opportunities to generate positive economic outcomes from cover crops including grazing and harvesting for livestock, reducing herbicide and fertilizer use, and sequestering carbon. Reduced soil erosion, improved water quality, and increased organic matter in soils and other services from cover crops have not been monetized but are important to protect soil and water resources and to improve soil fertility.

Dobberstein (2017) showed that U.S. farmers are moving away from intensive tillage while no-till, reduced tillage, and cover crop practices continue to grow. The census reported 15.3 million acres of cover crops were seeded in 2017 for an increase of 49% over the 2012 total of 10.2 million acres. The on-farm average of cover crop acres increased from 77 acres in 2012 to 100 acres in 2017 or 30%, so the benefits of cover crops are being recognized and applied.

What are Cover Crops?

We have already discussed some cover crops in Chapter 4 and their application in Chapter 5 under Multiple Cropping. Cover crops can be grass (corn, wheat, etc.) or broadleaf crops such as legumes. For example, in Iowa winter ground cover can be established as the soybean crop matures by aerially seeding one bushel of oats per acre with 20 pounds of either hairy vetch or rye. Cover crops can also be seeded into corn at the time of its last cultivation. Hairy vetch has been a good performer by stabilizing the soil and fixing nitrogen for future crops. The green manure crop in the 3-year rotation is grown for nitrogen fixation and protection of resources (National Research Council 1989).

Blanco (2023) has shown that cover crops have the potential to provide positive net returns if biomass production is sufficient. Cover crops can generate income which may offset the establishment costs and meet the original intended soil and environmental goals. Intercropping cover crops with other crops can meet economic goals and intended production as discussed above in Chapter 5.

Summary

The National Research Council in 1989 published a book called *Alternative Agriculture* to encourage United States farmers to consider alternative practices in agriculture. Alternative agriculture includes farming practices that attempt to reduce or eliminate chemical pesticides and antibiotics for pest and disease control, decrease soil erosion, reduce fertilizer use, and improve crop genetics to resist pest and disease problems.

Crop rotation was suggested to reduce pesticide use by growing leguminous forages and cover crops to disrupt weed, insect, and disease problems in rotation with corn, soybeans, and small grains. Fruit and vegetable growers can dramatically reduce pesticide use by integrated pest management programs. Manures can sometimes substitute for commercial fertilizer.

Because of genetic and cultural improvements and Roundup ready corn, sorghum, and soybean crops are being grown for alcohol production as well as food and feed crops. Cultivation of corn and soybeans is being done at record levels in the U.S. following the biofuels boom of the 2000s. Crop expansion has resulted in substantial conversion of long-term unimproved grasslands and nonagricultural lands to crops. Although the United States can produce a sustainable supply of biomass sufficient to displace 30% of the country's present petroleum consumption, oil remains the most feasible energy source.

Many woody and herbaceous plants can be used to produce alcohol but much farmland is required to grow the plant material. Biofuels require more energy to produce than the fuel generates.

Multiple cropping is growing and harvesting more than one crop on the same piece of land in a year. It has been practiced since antiquity but has only recently received scientific attention. Multiple cropping can be done simultaneously in mixture (intercropping) or of growing individual crops in sequence (sequential cropping). In the southeastern U.S., double cropping soybeans after wheat or barley is successful. In Texas (U.S.) cotton and sorghum have been grown in alternating four row blocks perpendicular to prevailing wind to provide protection of the young cotton plants from wind damage. The system is also used to protect young peanut plants. Many other examples are given of multiple cropping in the U.S., Tropical America, Asia, Africa, and the Middle East. Multiple cropping is usually practiced below 600 m elevation with an adequate water and growing season in the warmer regions of the world.

Chapter 5. Literature Cited

Holechek JL and Sawolhah MN (2014). Energy and Rangelands: A Perspective. Rangelands 36(6):36-43.

National Research Council (1989). Alternative Agriculture Board on Agriculture. Committee on the Role of Alternatives Farming Methods in Modern Production Agriculture, National Research Council, National Academy Press, Washington D.C. 448 p.

Smith, S, Diers B, Specht, Carver B editors (2014). Yield Grains in Major U.S. Field Crops. CSSA Special Publication 33, American Society of Agronomy, Inc. Crop Science Society of America, Inc. and Soil Science Society of American, Inc. Madison, WI, 478 p.

Lack TJ, Salmon M, Gibbs HK (2015). Copland expansion outpaces agricultural and biofuel policies in the United States. Environmental Research Letters 10044003 (http://iopscience.iop/1748-9326/10/4/044003) Accessed 3 April 2015.

Bovey RW (2016). *Mesquite: History, Growth, Biology, Uses and Management.* Texas A&M University Press, College Station, Texas. 261 p. (In Press).

Tittmann, P (2015). The Wood in the Forest: Why California needs to reexamine the role of biomass in climate policy. Pages 133-137 in Downing J, Thompson D, White H, eds. Woody Biomass: Energy, Ecosystems, Economics. California Agriculture 69(3):133-137.

Perlack RD, Wright LL, Turhollow AF, Graham RL, Stokes BJ, Erbach DC (2005). Biomass as feedstock for a bioenergy and bioproducts industry. The technical feasibility of a billion-ton annual supply. U.S. Department of Agriculture and U.S. Department of Energy. Oak Ridge National Laboratory, Oak Ridge, Tennessee. 60 p.

Nebel, BJ (1981). *Environmental Science: The Way The World Works.* 2nd Edition. Prentice-Hall Inc. Englewood Cliffs. New Jersey. 671 p.

Anonymous (2015). Ethanol Production Plants, Tribal Energy and Environmental Information Clearinghouse. Office of Indian Energy and Economic Development. Retrieved from teeic.indianaffairs.gov/er/biomass/restech/desc/ethanolplants/index.htm. Accessed 15 January 2016.

Anonymous (2016). Ethanol Facilities Capacity by State and Plant. Renewable Fuels Association, Washington, D.C. Nebraska Energy Office, Lincoln, Nebraska. Retrieved from www.neo.ne.gov/statshtm/122.htm. Accessed 15 January 2016.

Andrews DJ and Kassam AH (1976). The Importance of Multiple Cropping in Increasing World Food Supplies. Pages 1-1 *in Stelly* M, Kral DM, Eisele LC, Nauseef JH eds. Multiple Cropping, ASA Special Publication Number 27. American Society of Agronomy. Madison, WI.

Lewis WM and Phillips JA (1976). Double Cropping in the Eastern United States. Pages 41-50 in Stelly M, Kral DM, Eisele LC, Nauseef JH eds. Multiple Cropping, ASA Special Publication Number 27. American Society of Agronomy. Madison, WI.

Gomm FB, Sneva FA, and Lorenz RJ (1976). Multiple Cropping in the Western United States. Pages 103-115 in Stelly M, Kral DM, Eisele LC, Nauseef JH eds. Multiple Cropping, ASA Special Publication Number 27. American Society of Agronomy. Madison, WI.

Pinchinat AM, Soria J, Bazan R (1976). Multiple Cropping in Tropical America. Pages 51-61 in Stelly M, Kral DM, Eisele LC, Nauseef JH eds. ASA Special Publication Number 27. American Society of Agronomy. Madison, WI.

Harwood RR and Price EC (1976). Multiple Cropping in Tropical Asia. Pages 11-40 in Stelly M, Kral DM, Eisele LC, Nauseef JH eds. Multiple Cropping, ASA Special Publication Number 27. American Society of Agronomy. Madison, WI.

Okigbo BN and Greenland DJ (1976). Intercropping Systems in Tropical Africa. Pages 63-101 in Stelly M, Kral DM, Eisele LC, Nauseef JH eds. ASA Special Publication Number 27. Multiple Cropping. American Society of Agronomy. Madison, WI.

Nasr HG (1976). Multiple cropping in some countries of the Middle East. Pages 117-127 *in* Stelly M, Kral DM, Eisele LC, Nauseef JH eds. ASA Special Publication Number 27. American Society of Agronomy. Madison, WI.

Blanco H (2023). Economics of Cover Crops. Crops and Soils, 56 (5) 4-12.

Dobberstein J (2017). "U.S. No-Tilled Acres Rises 89% in U.S. Intensive Tillage Sees Major Decline." Retrieved from https:// no.tillfarmer.com/articles/8683-no-till-acres-rises-in-us-intensive-tillage-sees-drastic-decline. Accessed 14 April 2023.

HORTICULTURE (GARDEN) AND FLORICULTURE

Introduction

Horticulture is the branch of agriculture that deals with the art, science, technology, and business of growing plants (Anonymous 2016). It includes the cultivation of medicinal plants, fruits, vegetables, nuts, seeds, herbs, sprouts, mushrooms, algae, flowers, seaweeds, and nonfood crops such as grass and ornamental trees and plants. It also includes plant conservation, landscape restoration, landscape and garden design, construction and maintenance, and arboriculture. Inside agriculture, horticulture contrasts with extensive field farming and animal husbandry.

The word horticulture is derived from *hortus* (garden) and *cultura* (culture or cultivation), hence the culture or cultivation of garden crops. The word agriculture, derived from *agri* (field) and cultura, refers to the cultivation of field crops. The term agriculture also includes animal husbandry, poultry husbandry, dairy husbandry, dairy manufacture, and others. Likewise, the term horticulture now means more than care of gardens (Shoemaker 1952).

Value of Horticulture in the U.S.

Horticulture is extremely popular in the United States and world. It produces high-quality food and aesthetic pleasure. Horticulture and agronomic programs are offered at land-grant universities in the U.S. with associate, baccalaureate, master's and doctoral degrees. Jobs in horticulture are available, and possessing a degree helps in obtaining good jobs and business opportunities (AgEdLibrary.com date unknown).

Floriculture is an international multibillion-dollar industry and includes the growing, distribution, and processing of flowering and foliage plants. Floral crops were estimated at $5.2 billion in the U.S. in 2004 on the wholesale market. California, Florida, Michigan, Texas, and New York lead the United States in production (AgEdLibrary.com date unknown).

Landscape horticulture value was estimated to be $3.3 billion in 2000 by the USDA that included deciduous shrubs and trees, broadleaf and coniferous evergreens, fruit and nut plants and other plants. Homeowners in the United States spent over $50 billion to install, improve, and maintain their landscapes and gardens as determined in 1994 by the Gallup Organization (AgEdLibrary.com date unknown). In 2004 vegetables made up more than $10 billion of U.S. farm receipts; and nut crops, citrus, and noncitrus fruit production garnered 2.4, 3, and 9 billion dollars, respectively (AgEdLibrary.com date unknown).

Examples of Plants Grown

Vegetables

Vegetables grown for the processing industry include tomatoes, peas, sweet corn, snap beans, cucumbers, carrots, lima beans, and spinach. Vegetables for the local market may contain potato, sweet potato, tomato, sweet corn, cabbage, lettuce, onion, cucumber, carrot, spinach, asparagus, and muskmelon. Vegetable forcing in greenhouses and mushrooms and special structures for rhubarb include tomato and cucumber, mushroom and rhubarb (Shoemaker 1952).

Tree and Perennial Crops

Citrus includes orange, grapefruit, lemon, and lime. Deciduous fruit trees are apple, peach, plum and prune, pear, and cherry. Nut and tender fruits include almond, filbert, pecan, walnut, fig, olive, avocado, and date (date palm). Small fruits include grape, strawberry, raspberry and blackberry, blueberry, currant, and gooseberry.

Flowers

Cut flowers include rose, carnation, chrysanthemum, gardenia, and orchid.

Flowering potted plants are poinsettia, cyclamen, begonia, hydrangea, saintpaulia, and orchids. Bulb plants imported are tulip. narcissus, hyacinth, crocus, and lily.

Foliage potted plants include philodendron, ferns, and palms. Charts are available for varieties, frost free days, and planting dates for a given location from the horticulture industry, nurseries, publications, and local state agricultural universities.

Other Horticulture Practices

Other horticulture practices include the nursery industry that supplies all types of plants to homeowners and business locations, the turf industry, and landscape gardening and design industry.

We will discuss seed production, the food and aesthetic value of horticulture crops and plants as well as their environmental needs for growth as we progress through this chapter.

Basic Knowledge of Gardening

The theme of this book is to introduce or review where our food comes from and how it is grown or produced. No one knows better than the accomplished gardener how to grow food and prepare or preserve it for human consumption or for aesthetics. There is a real sense of accomplishment, but time-consuming, to grow garden plants but hopefully rewarding in high quality produce and satisfaction.

Knowledge of tools and equipment needed, best time to plant, adapted plant varieties, soil and pest management are some requirements. Gardening and food preservation assistance can be obtained from the Cooperative Extension Service of the U.S. Department of Agriculture and/or the land-grant university system. There are home economists and agricultural extension agents or farm advisors in most counties in the U.S. Consultation can be online, by phone, bulletins, circulars, facts sheets, newsletters, and books. Publications are usually free except for the more elaborate ones. USDA publications are also available from the U.S. Government Printing Office, Washington DC 20402, or some members of Congress.

Information on gardening is also available from neighbors that garden, local garden clubs and commercial florists, greenhouse and nurseries, local public libraries, courses at local universities, and consultation with local garden centers and seedsmen. Soil testing is usually available from the land-grant university, county extension office, or commercial testing laboratory to determine nutrient requirements and fertilizer needs for a specific crop grown.

Americans garden today because of increasing food costs, ability to produce fresh and quality food, personal satisfaction, therapy and exercise, family or neighbor contact, artistic and aesthetic value, and many other reasons.

Vegetable Gardening

Preparation

Whether personal gardening or commercial vegetable growing, specific requirements are needed to be successful. For small and commercial sites, important factors include sunlight, good air circulation, protection from herbivorous animals and competitive vegetation, good soil structure, and fertility.

Most fruit trees and vegetable plants require direct sunlight most of the day. Some leafy vegetables will tolerate partial shade, but fruit trees and fruit-producing vegetables need direct sunlight.

Air circulation is important in reducing plant diseases, especially in humid areas.

Prolonged damp foliage following heavy dew or rainfall favors development and spread of fungus or bacteria diseases. Too much shade adds to the problem. In the commercial setting, windbreaks may be needed to protect plants from strong prevailing winds such as the U.S. Great Plains.

Protection from animals is often necessary with a close woven fence against rabbits, squirrels, dogs, and certain wildlife. Avoid competition from trees and large shrubs as far as possible. In addition, shade tree or shrub roots may extend into the garden area to rob moisture and nutrients from the crop. Cutting woody plant roots and controlling weeds may be necessary for full crop potential.

Good soil drainage, structure, and fertility ensure good crop growth. A loamy, fertile soil with a minimum of clay subsoil is desired. Structure and fertility can be improved by addition of organic matter in heavy clay soils. Compost or manure, as well as lime and fertilizer, will help. Soil testing will indicate the need for lime and fertilizer.

Good soil drainage is also important in providing air to the crop roots. Organic matter improves the moisture-air balance in moist soils. If the selected site floods and is poorly drained, drain tile, drain ditches, or breaking up the subsoil may help (Blackwell 1977).

Selecting Varieties

Growing the best varieties for your area is highly important. Varieties adapted to certain soils and climate are best. Plant breeders consistently develop new varieties and many disease-resistant varieties are available. Varieties that tolerate certain recommended pesticides are desired. It is generally wise to purchase new seed each year. However, last year's seed of some vegetables can be stored in a cool, dry place free of rodents and insects. Seed of sweet corn, leeks, okra, onion, parsley, parsnip, rhubarb, and salsify should be replaced each year. Saving seed from your own garden should probably be avoided as many new varieties are hybrids and resulting plants may deviate from expected characteristics (Blackwell 1977).

When to Plant

The general rule, the more cold-hardy, cool-season vegetables should be planted or transplanted 4 to 6 weeks before the last spring frost. Less hardy cool-season crops should be planted 2 to 4 weeks before the last spring frost. Cold-hardy crops include broccoli, cabbage, cauliflower, collard, kale, kohlrabi, lettuce, onions, English peas, Irish potatoes, spinach, and turnips. The less-hardy cool-season crops include beets, carrots, chard, mustard, parsnips, and radishes.

In most areas of the country, frost-tender, warm-season, or heat-tolerant vegetables should not be planted until after the last spring frost and in warm soil. They include New Zealand spinach, squash, sweet corn, and tomato transplants. The heat-tolerant crops such as lima beans, sweet potatoes, cucumbers, cantaloupes, pumpkins, watermelons, eggplant, peppers, okra, Southern peas should not be planted until a week or two after the first frost-free date (Blackwell 1977).

Some cold-hardy, cool-season vegetables can be planted in late summer or early fall and mature during cool weather. Dates for planting the various vegetables are available locally.

Cultivation and Weed Control

Cultivation is done to control weeds. Destroy weeds when they are small, otherwise they become too large and difficult to deal with. Cultivation also loosens soil for better aeration. Mulching also controls weeds and conserves moisture. Sawdust, leaves, ground-up bark, rotting hay, straw, or lawn clippings work. Black plastic sheeting can also be used to control weeds and conserve moisture but may cause high soil temperature in hot summer. It is best used in early planted crops (Blackwell 1977).

Record Keeping

Keep a record of the crops, when planted, and any problems that develop. This will help next year in avoiding problems. This may help in determining if you planted too much or too little of a desired vegetable. Keep track of costs for seed, plants, fertilizers, pesticides, supplies and equipment, and the worth of your labor and compare it to actual costs at the grocery store. This may help you decide future direction.

Location of Garden

The garden site should, as discussed earlier, avoid too much shade, shrub, and tree roots. Site selection for fruit or nut trees is more critical than for vegetables because fruit or nut trees cannot be moved around like vegetable or flower gardens. Most professional landscapers do not prefer to integrate fruit or nut trees in a landscape but would rather put them on the back side of the property, with the exception of pecan and oriental chestnut trees (Wilson 1977).

Water availability is an important consideration, so locate the garden near a water source. Vegetables and fruit trees grow slower and mature later if shaded. Shade not only reduces light but may reduce soil temperature and delay growth and maturity. Gardens should also be protected from strong winds by trees or shrubs or some other structure. Some crops should be planted near south- or west-facing walls where plants can benefit from the heat and reflected light. Frost may also occur in low areas while the hillside allows cold air to drain and warm air to rise past the garden site. As indicated, poorly drained soil can be improved by adding organic matter and raising the level of the beds above the surrounding ground or installing drainage tile (Wilson 1977).

Container Gardens

Container gardens can be found where larger gardens are not possible for apartment and mobile home residents. Fruits, berries, and vegetables can be grown in small containers. They can be very decorative and practical. To reduce the frequency of watering, containers should be a minimum of 4-gallon capacity for vegetables. Tomatoes (two plants per container) can be planted in a 20- to 30-gallon plastic garbage can. Berry and fruit tree plantings require sturdier containers to support root weight.

Window boxes can be used to grow strawberries and small leaf or root vegetables. Midget varieties of cucumbers, tomatoes, and melons can be grown in window boxes in about 1.5 cubic feet of soil for each plant (Wilson 1977).

Front Yard Garden

Vegetables can be grown in front yard gardens, such as fresh peas, carrots, tomatoes, and melons, but are usually poor locations for fruit, nut, citrus, or bush fruits because of possible vandalism and site obstruction (Wilson 1977).

Community Gardens

Most successful programs have been organized by volunteer committees including experienced gardeners. Someone needs to be in charge to assign plots, collect fees, charge out tools, and provide advice and arbitrate disputes. Fees should cover water bills, tillage, fence lines, supplies, maintenance, and a stipend for supervisors. A high, stout fence with locked gates is usually required (Wilson 1977).

Garden Tools

Basic tools needed for small gardening are a spading fork, steel rake, and garden hoe (Bartok 1977). Additional hand tools are a round pointed shovel, a handheld or wheel-supported cultivator. Hand forks or claws are miniature cultivators that are useful in a small garden.

Many types of seeders are available, but we did all seeding by hand in the garden as a young boy after the soil was prepared. Most seeders are adapted to a variety of seed sizes. Some seeders can do it all - open the furrow, drop and cover the seed, and firm the soil in one operation.

Lime and fertilizers can be applied by hand or by use of spreaders which are probably more accurate. A wheelbarrow or cart helps in moving soil, stones, peat moss, fertilizer bags, and tools. Rotary tillers with steel tines can be used to prepare a seedbed. Compact or garden tractors can be equipped with several implements that make gardening easier by a mold-board plow, disk, spike-tooth or spring-tooth harrows. Garden tractors can also pull rotary tillers, carts, fertilizer spreaders, seeders, and sprayers (Bartok 1977).

Pest Management

Pest management should be recognized as a goal of the successful gardener. Removing unwanted weeds is probably the biggest problem. Weeds need to be removed early and when they are small before they cause damage and are easy to remove. This can be done by hand pulling, hoeing, cultivation, and sometimes spraying or pellet application of herbicides.

Control of harmful insects and plant diseases are other problems to be aware of. Learn what to expect by getting literature on the various vegetables and fruits grown in your area. Informative pamphlets and bulletins or online information can be obtained from federal and state agricultural departments and cooperative extension services. These extension offices are usually located at county seats and are affiliated with state colleges and universities where agriculture is practiced. Newspapers with a gardening section are useful as well as radio, your local garden centers, or chemical supplier.

Rodents, rabbits, birds, and deer can also be problems. Rodents can be trapped. Rabbits can be repelled with dried blood meal or fencing. Birds can be kept off fruit trees by nets, scarecrows, or noisemakers. Deer can be repelled with electric fences, dried blood meal, mothballs, and other means (Scheel 1977).

Organic Gardening

Organic gardening is the production of crops without the use of chemical fertilizers or pesticides (Judkins 1977). However, manure, sewage, sludge, cottonseed meal, bone meal, or dried blood can be used for nutrients. Weeds, diseases, and insects are controlled by natural resistance, birds, predators, insects, or mechanical means rather than chemical.

One of the most important uses of organic matter by the home gardener is as a mulch. Mulches reduce rainfall and irrigation water-erosion, increase water infiltration, and conserve soil moisture. Leaves, lawn chippings, fresh sawdust, fine wood shavings, pine needles, chopped straw, ground corn cobs, and other organic materials make good mulch.

Cover crops, such as planting of rye or wheat in the fall, serve to control soil erosion and provide organic matter for next year's garden crop. If crops are still growing in the garden, the rye or wheat may be planted between the rows. The cover crop can be plowed or rototilled into the soil in the spring when the garden is prepared for planting. If weeds are present (two or more inches tall), kill them by cultivation or hoeing before the mulch is applied. Mulch will usually smother weeds less than one inch tall.

Composted humus material is especially useful in making a soil mixture for producing seedling plants for the vegetable garden. It is also useful as a mulch.

Although some vegetables may be severely damaged by diseases or insects, many can be grown successfully without use of pesticides. There are also nonhazardous organic or biodegradable pesticides available.

Biocontrol utilizes lady beetles to feed on aphids. Cutworm damage can be reduced by a cardboard collar encircling the plant a half inch out from the stem, extending an inch into the ground and 2 inches above. A ring of wood ashes or sharp sand around plants helps control slugs. *Bacillus thuringiensis* is a bacterial disease effective against the larvae of a number of moths and butterflies. It is useful for controlling cabbage loopers on broccoli, brussel sprouts, cabbage, cauliflower, collards, and kale. Hybrid sweet corn varieties can be selected for resistance to corn borers. Other means are available to control insects in beans, pumpkins and squash vines, cucumbers, eggplant, potato, and tomato varieties.

Plant Growth and Reproduction

Fruit and vegetable plants are made up of tiny cells. They grow and reproduce by number and size. Seeds germinate and become young plants that develop roots, stems, leaves, and flowers. When the flowers are pollinated, seeds and fruit form (Combs 1977). This process was explained in detail in Chapter 4 on agronomic crops for corn and other field crop plants.

These processes involve uptake of carbon dioxide and oxygen, largely from the air and water, and minerals from the soil. This growth process results in development of fruits, seeds, stems, tubers, leaves, and flowers that provide nourishment and aesthetics for people and animals.

Plants photosynthesize from energy from the sun with carbon dioxide and water to produce simple sugars, and oxygen is released into the atmosphere. Complex carbohydrates, fats, proteins, and vitamins are formed by the incorporation of sugars with mineral elements from the soil. As stated in Chapter 1, agriculture and gardening are the only major sections utilizing solar energy on a broad scale. Therefore, all kinds of plants produce wood, food, and beauty from components already in the environment. Humans direct some of these processes for shelter, clothing, food, feed, beauty, and our well-being, but a large part is ongoing in nature.

Most vegetables are reproduced sexually from seeds. Most fruits and a few vegetables are reproduced asexually or vegetatively from plants or plant parts. Seeds contain the embryo (miniature live plant) with stored food for germination and early growth (endosperm). The seed coat must be broken for the tiny plant to emerge. A suitable temperature and moisture are required for germination.

The root must emerge and anchor the plant and begin to absorb water and minerals. Roots vary in spread and depth depending upon the crop. Roots such as beets, carrots, and sweet potatoes are major food storage organs and depend on leaves for their development. Leaves originate from stems and are the principal food-manufacturing organ for green plants (Combs 1977).

To produce seed, plants must first produce flowers which have four parts: sepals, petals, stamens, and ovaries if complete. The sepals are the outermost parts and together form the calyx. It serves as a protective cover on a bud with the petals, stamens, and ovaries inside. Next inside are the stamens which furnish pollen, and in the center of the flower are one or more ovaries. The ovaries contain ovules which develop into seeds when fertilized with pollen. Birds, insects, wind, and water are important pollen carriers.

Once the pollen grain is on the stamen, it germinates to form a tube which grows through the stigmatic surface and down through an often elongated column of tissue called a style into the ovary. The pollen tube fuses with the egg nucleus of the ovule to form the seed embryo (zygote). Another sperm nucleus fuses with other nuclei of the ovule to form the seed endosperm.

The seed is made up of an embryo and one or more seed coats and usually an endosperm (see Chapter 4). The embryo (tiny miniature plant) usually has a plumule or rudimentary shoot; one or more cotyledon or seed leaves; a radicle or rudimentary root and a hypocotyl, the part of the embryo between radicle and cotyledons. The seed coat protects and encloses these parts. The endosperm and cotyledons provide stored food for germination and growth. The cotyledons may act as leaves to carry on photosynthesis and manufacturing of food until the first true leaves support the new seedling (Combs 1977).

Asexual or vegetative reproduction occurs through cell division. In asexual reproduction a cell can split the chromosomes and divide into two daughter cells. As a result, the complete chromosome system of the cell is duplicated in each daughter cell. Asexual reproduction is important because unique characteristics of an individual plant can be maintained.

Cuttings to produce new roots and shoots (plants) can be sometimes done from stems, leaves, and roots from the parent plant. Grafting of one part of a plant to another is sometimes used. Tubers (potatoes), bulbs (onions), and tuberous roots (sweet potatoes) can also be used to produce new plants by cutting or pulling the thickened structures into pieces and planting them. Apples and other fruit trees are produced in great numbers by planting seed for root stocks and then grafting or budding the desired scion wood or variety to the root stock (Combs 1977).

We discussed the importance of pollination in Chapter 4 for sexual reproduction of the flower for seed and fruit development. The honey bee is most effective of all pollinating insects. There are a large number of fruits and vegetables that benefit from insect pollination. Some common fruits and vegetables include apple, apricot, blackberry, cherry, lima bean, peach, pear, raspberry, strawberry, and watermelon. A few plants will produce fruit without any form of pollination, but the fruit is seedless and is called parthenocarpic. They include seedless oranges, raisin grapes, certain cucumbers, certain pears, pineapples, some figs, and bananas.

Fruits and Nut Crops

We have touched lightly on fruit and nut crops, but they need to be discussed separately because of their special requirements. They require full or nearly full sunlight and good drainage. Soil should be deep and rich, ideally three to four feet for fruit and six feet or more for pecans with good air circulation and available water (Sperry 1982).

Fruit trees should be planted during winter when the tree is dormant in Texas. Planting in colder areas may be best in the fall. In extremely cold areas, spring planting may be best (Shoemaker 1952). They are usually sold bare rooted and winter planting gives them time to establish new root growth before spring. Proper spacing depends on the kind of plant, but fruit trees require adequate space. Apples and pears require a spacing of 25 to 30 feet between plants whereas recommended spacing for apricot, cherry, citrus, figs, grapes, muscadines, peaches, persimmons, and plums are less than 25 to 30 feet. Pecan trees require spacings of 35 to 45 feet between plants (Sperry 1982).

Not all fruit trees are self-fertile so a second tree is required of a different variety to produce pollen. This may include apples, apricots, blueberries, pears, pecans, plums, and walnuts.

Fruit crops may take several years to produce fruit and need sufficient vegetative growth before they can reproduce. Sperry (1982) reported that using adapted varieties for the area, planting vigorous and healthy varieties in well drained fertile soil with proper care should hasten fruit production.

Most deciduous fruit crops have a need for exposure to cold before it can bud and bloom in the spring. Chilling requirements of the fruit crop vary, even within the same type of fruit. Peaches with as little as 250 hours of chilling do well in South Texas, whereas those of chilling requirements of 800 to 1000 hours are better adapted to Central and North Texas. Exposed to cold in Texas is measured by the fruit crop exposure to cold below 45°F and above freezing (Sperry 1982).

Care of fruit crops start and end with proper planting, pruning, fertilizing, watering, and controlling insects and diseases and harvesting (Shoemaker 1952; Sperry 1982).

Floriculture

As stated above, floriculture is an international multibillion-dollar industry and is the growing, distribution, and processing of flowering and foliage plants. The heart of all commercial floriculture crop production is plant propagation. The decision to propagate plant species commercially must be based on economics, quality, and dependable delivery (Dole and Wilkins 2005).

Propagation

The propagation methods have been discussed and can be separated into sexual methods - seeds and spores and asexual methods of cuttings, layer divisions, natural reproduction structures such as bulbs, grafting and in vitro micropropagation. Many bedding plants and a significant number of cut flowers, potted flowering plants, and foliage plant species are propagated from seeds. In a number of species, however, sexual reproduction produces plants too variable for commercial production so asexual methods are used to produce clones (individual plants propagated from one original plant). The clones have the same traits as the original plant (Dole and Wilkins 2005).

Cultivar Protection

Many countries, including the United States, now allow patent protection for genetically-modified (GM) plants and seeds (Anonymous 2012). Cultivation of GM plants has been widely adopted, mainly because of convenience and cost-effectiveness.

GM plants have been grown commercially since 1966 reaching 160 million hectares (39.5 million acres) by 2011 worldwide. This amounts to more than 7% of the total cultivated land area of the earth. GM soybeans with herbicide resistance and insect-resistant corn and cotton are widely used in the U.S.

Dole and Wilkins (2005) stated that plant patents can be obtained on asexually propagated plant materials that are new and unique. Patents are effective for 20 years, after which the material can be freely propagated and marketed. Plant patents are issued in the United States by the U.S. Patent and Trademark Office. Plant patents are independent of trademarks. Trademarks can be words, symbols, or designs to identify the goods and services of a party. Trademarks can be combined with patents but trademarks do not expire.

Plant variety protection can be for inbred plant varieties produced from seed. Plant utility patents are used by biotechnology companies to protect unique production processors, genes, plant parts, and physical traits. Trade secrets are used to protect parent inbred lines in the production of F1 hybrid seed propagated cultivars. Proprietary rights can be used to control the propagation and marketing of a unique plant without patents, trademarks, or variety protection. Distribution of the plant material is controlled by contractual agreements, making propagation and marketing of a plant by anyone other than those specified in the contract illegal. Many countries other than the United States also have legislation to protect the work of individuals and companies developing new cultivars.

Growing Flower and Foliage Plants

Whether growing floriculture crops around homes and business locations or commercially in the greenhouse or field, basic requirements similar to growing vegetables are needed.

If plants are grown from seed, specific conditions for each species may be required for germination. This includes proper lighting, germination temperature, and germination time. It also includes uptake of water by the seed to start the germination process.

The seeds are living plants and must be handled correctly to germinate. The seed must be properly stored until use and tested for germination before planting. Some seeds require scarification (breaking a hard seed coat) before they will germinate and grow. This can be done mechanically, soaked in growth-enhancing chemicals, hydrated in hot water, or primed in salt solution or polyethylene glycol to restrict the rate of water uptake and provide a more uniform product. Some seeds require a dormant period before germination. This and other seed germination, as well as propagation requirements for a large number of species, have been provided (Dole and Wilson 2005).

To grow plants after germination in the greenhouse selected containers, growth media, temperature and water regimes, proper nutrition, and disease control must be administered. This also applies to plants propagated by asexual such as stem, root, and leaf cuttings.

The flowering process is an especially important subject in floriculture. The three main environmental control stimuli include the proper photoperiod, light intensity, and temperature. Each plant species has different environmental requirements for flowering. First the plant must be mature before the flowering stimulus is present. This occurs after the juvenile period. The juvenile period in floriculture is usually relatively short, and plants need a minimum number of leaves to reach maturity. Plant mass or size can also be important.

Plant growth is controlled by naturally occurring substances in the plant. They include auxins, gibberellins, cytokines, ethylene, and abscisic acid. Commercially, auxins are used to enhance root development during propagation and tissue culture. Gibberellins are involved in shoot elongation, flower development, and seed germination. Cytokines are important in branching, cell division, and juvenility. They are used in tissue culture to stimulate shoot growth and delay leaf aging during postharvest. Ethylene or ethylene-producing chemicals are used commercially for height control, branching, leaf and flower abscission, flower induction, and flower inhibition. Anti-ethylene agents are widely used to prolong postharvest life and delay flower, petal, and leaf senescence (aging). Abscisic acid promotes senescence and directs the storage of photosynthates made by the leaves. Currently there are no commercial uses for abscisic acid in floriculture (Dole and Wilkins 2005).

Pest management begins with pathogen-free media, containers, benches, and water systems in the greenhouse. During production a pathogen-free environment, pest-resistant plants, proper irrigation and humidity is needed. If problems occur, identify the problem, use the appropriate control method and rotate methods, and remove diseased or heavily insect-infested plants immediately.

A floriculture crop is at its highest quality at harvest and must be properly handled to minimize loss of quality. To maintain market quality plants and cut materials, plants must be handled at the correct temperature and high carbohydrate levels for best postharvest life. Growing plants in the greenhouse, growth chambers, nursery, and field is a very complex and important industry in the U.S. and world. Marketing and growing quality crops are paramount. The industry brings us great beauty and joy from the many varied flowers and plants available. The reader is referred to *Floriculture: Principles and Species* by Dole and Wilkins (2005) or any floriculture or horticulture reference for further details on growth, harvesting, postharvest, and marketing requirements for each plant species.

Landscaping and Turf

Since this book is about food production, there is still a place for a beautiful landscape around homes and commercial establishments with use of vegetable, fruit, and nut crops and with other shade trees, flowering plants, and shrubs. Turf grass or some ground cover (English ivy) may not add any food value, but it adds significant beauty and soil stability to the landscape.

Commercial landscape gardening is an important part of horticulture. To get the most benefit from the landscape, good planning is required. The grounds around a house should consider: 1) foundation planting, 2) lawns, 3) boundary planting, screens and hedges, 4) work areas and service yards, 5) vegetable and fruit areas and 6) special features such as rock gardens, rose gardens, and pools (Shoemaker 1952; Sperry 1982).

Most homeowners want a good lawn around their home. Turfgrasses reduce dust, glare, and surface temperatures and provide playing surfaces for sports and athletics. In addition, turfgrasses prevent erosion and add beauty on golf courses, playgrounds, parks, roadsides, ditch banks, mine spoils, sanitation fills, and other areas (Duble 1996).

Greenhouse Use

Growing flowers, vegetables, and young shrubs and trees in the greenhouse is big business in the U.S. and worldwide. The greenhouse use is responsible for much of this activity, especially flowers and starting young plants.

Boodley (1998) stated that the top six potted florist plants from first to last include potted foliage, poinsettia, chrysanthemum, azalea, Easter lily, and African violet. The top six florist crops sold from first to last are bedding/garden plants, potted foliage, potted poinsettias, roses (hybrid teas), and foliage hanging baskets.

Weather is a factor in the location of a flower growing business. In the United States California, Florida, Texas, Hawaii, and Colorado are big flower and potted foliage producers.

Greenhouse design varies with needs from small gardeners to large commercial settings. Greenhouses complement nursery and field sites to complete the growth and life cycle of many plants. Specific growing conditions exist for each type of plant grown, and providing the proper environment is necessary and expensive.

Vegetable Seed Production

Thompson (2008) pointed out that seeds sustain us not only in foods but also in fabrics, detergents, and fuels. Seeds are vital to farming, vegetable gardening, and raising livestock. All life depends directly or indirectly on crop seeds. Large-scale agribusiness underpins modern economies and the consumer lifestyles that they provide.

George (1999) stated that cultivated vegetable species were developed by selection from local wild plants and subsequently supplemented by plant introductions from other areas and later still from other continents. Vegetables are part of a balanced diet and inclusion of vegetables in cereal-based systems and year-round vegetable production systems. Successful vegetable production is highly dependent upon the sustainable seed supply.

Vegetables are an essential part of the diet and supply fiber, trace minerals, vitamins, folacin, carbohydrates, and protein. Vegetables supplement staple foods such as rice, corn, or wheat and use of canning, freezing, and dehydration techniques enable a more diverse range of vegetable types and quality food.

Two basic ways are open for farmers or individuals to ensure seed for next year's crop. One is to save seed from the grower's crop or purchase seed from the government or a private company. In developing countries, a majority of vegetable seed is saved by farmers and growers from their crop. In developed countries, a high percentage of seed is purchased from private companies.

Most vegetable crops are grown from seed and only a small number are propagated vegetatively like cassava, garlic, Jerusalem artichokes, sweet potatoes, and yams.

Plant Breeding

Plant breeders are protected when a cultivar is developed by the Convention of the International Union for the Protection of New Varieties of Plants (George 1999). The main criteria for the cultivar is that it is distinct from existing cultivars and has not been commercialized before.

The pollination, fertilization, and formation of the seed has been discussed as well as the importance of insects to pollinate vegetable and agronomic crops.

Hybrid cultivars are also used in vegetable seed production. The most widely used is the F1 hybrids by vegetable breeders. An F1 hybrid is produced by crossing two distinct lines. In practice the two parent lines are the result of inbreeding (the production of offspring by fusion of genetically closely related gametes). Commercial seed of F1 hybrid cultivars are the result of crossing two inbred lines which are maintained under close supervision, which are known to produce a desirable hybrid.

The advantages of F1 hybrid cultivars include uniformity, increased vigor, earliness, higher yield, and resistance to specific pests and pathogens (George 1999). F1 hybrid seed lots are more expensive compared to open pollinated cultivars because of the breeding program, subsequent maintenance of inbred parents, extra land for male parents, care with planting, isolation and harvesting, high labor costs for hand emasculation (breeding), and sometimes lower seed production. The reader is referred to other sources for more details on breeding techniques.

Growing Seed

The principles and practices to establish the crop are the same as for the production of vegetables for market outlets (George 1999). However, it is important that care be taken to avoid admixture of seeds or plant material at all stages so the genetic quality or "trueness to type" can be fully maintained.

Harvesting and Processing

Once mature there are losses before harvest by seed ripening and dropping off before later seed has developed. Losses before harvest occur because of "shattering" or shedding of seed.

Other sources of seed loss before harvest include bird and rodent feeding and inclement weather. Shattering may be reduced by polyvinyl acetate sprays to glue the seed in place before shattering. Bird losses can be prevented by scarecrows, aerial balloons, noisemakers, rattles, and other devices. Netting is successful for the protection of small-scale areas but impractical over large areas.

Lodging (seed plants fall to the ground before harvest because of the weight of the plant or seed heads) can cause deterioration and seed quality loss. Lodging can be caused by wind, heavy rainfall, and top-heavy plants from overfertilization.

Harvesting of seed should be done at peak time before shedding, during proper ripening and moisture content for best quality and germination potential. Determining best harvest time is difficult and sometimes impossible to do because of workloads and inclement weather. Each species differs for the best harvest method in time. Hand harvesting is still done for very high value crops (George 1999).

Storage of Seed

After vegetable seeds are extracted from the greenhouse or field and have been properly dried to 12% or less (9% for oily crops), it can be stored. The moisture level for seeds to be stored in sealed containers should be less than these figures.

The storage period may be for a few weeks to a few years. The need to store and safeguard seed stocks may be for immediate use after natural or man-made disasters. This may include drought, flooding, war, or other disasters that can result in sudden seed shortages. This is true for agricultural staple crops such as rice, wheat, and corn as well as vegetable seed (George 1999).

Satisfactory seed preservation and storage assists the industry in buffering the variations in seed requirements. The storage of seed as germplasm, a potentially valuable genetic resource, may require a very long storage time in small seed lots. Germplasm centers (gene banks) use specialized storage methods for long-term storage of seeds. Some species have short-lived longevity in storage in spite of storage conditions.

Good nutrition before seed harvest and favorable environment favor seed longevity, storage, and viability. Seed free of harvesting damage and storage pests and pathogens extends longevity and seed quality. After harvest, most seed suitable for storage has a moisture content not greater than 10% of seed weight. At this moisture level, metabolic rate is extremely low; but it is usually capable of taking in water when planted in a moist soil for germination and growth. Some species require a dormancy period before they will germinate.

Ideally each seed lot should be stored at a specified temperature with proper ventilation and humidity. Seed should be protected from destructive insects and rodents. Seed should be packaged and labeled for identification for sale and regularly inspected for any deterioration or damage. Testing procedures are available to determine seed viability.

Home and Commercial Food Presentation

The variety of preserved foods, including fruits and vegetables, that are not only available fresh but also preserved in plastic packaging or metal cans is almost beyond comprehension in the grocery store. These packaged and canned commercial preparations are widely available in the U.S. and many other countries and provide an instant food source with little preparation by the consumer. Most Americans in the U.S. live miles from farmland or garden areas and may have no interest or time to grow or preserve their own food.

Cause of Spoilage

Historically, food preservation and processing was done to assure a food supply to prevent starvation. Since raw foods are living biological entities from animals and plants, they also provide a source of food for microorganisms which can grow on or in them, spoiling food before it can be eaten or used. The primary objective of food preservation is to prevent food spoilage by preserving it for human and animal consumption. Food preservation techniques followed correctly prevent foodborne disease such as botulism or any form of food poisoning (Zottola and Wolf 1977).

A second chance of food spoilage is vermin such as rodents, rats, mice, and insects that attack food and eat or contaminate it before human use. Food can also become undesirable because of senescence (aging) and chemical deterioration making it rancid in taste, loss of color or bleached.

The last cause of food spoilage concerns food handling. Physical or mechanical damage to food causes bruising, crushing, cutting, and wilting or water loss. These mechanical defects deface appearance and allow easier entry of microorganisms, insects, and other vermin to cause spoilage and aging.

Packing and Preservation

Packaging is a convenient method of handling food, prevents contamination during and after processing, prevents vermin infestation, supplies a container for storage, and is a necessary part of preservation. A mason jar properly sealed for pickling is an example.

The major method used for home preservation of food is temperature control. This includes canning with the pressure canner or a boiling water bath, blanching food before freezing it, refrigerating food, or freezing it. By increasing the temperature of food, microorganisms are destroyed. When the temperature is decreased, like refrigeration or freezing, their growth is inhibited.

On the lower side of pH 4.6, acid content of food will prevent botulism (*Clostridium botulinum*) and most other spore-forming bacteria. Most common types of spoilage microorganisms associated with acid foods are yeast and molds. They are acid-tolerant and grow in acid environments. They are killed at a lower temperature than spore-forming bacteria. Acid foods only need a heat treatment in a boiling water bath for a specified time to destroy the microbes present.

Commercial canneries, which are regulated by U.S. Government laws, have little problem with botulism canned foods. Most problems are from home canned foods.

Preservation of food by controlling the acid content can be achieved by naturally fermenting the food, like turning cabbage into sauerkraut. Another method is to add an organic acid to the food to reduce the pH, like adding vinegar to cucumbers to make pickles (Zottola and Wolf 1977). Some foods, such as berries and fruits, naturally contain enough organic acid so their pH is below 4.6 and preservation of these foods requires only a boiling water bath heat treatment or freezing.

Drying food is one of the oldest preservation methods. Water removal can be by sun drying or oven drying. Homemade or commercial dehydrators are available to dry fruits and vegetables. A temperature of 135° to 140°F is desirable for a dehydrator or oven drying. Moisture must be removed as fast as possible at a temperature that does not affect the flavor, texture, color, or nutritive value of the food. After drying, food should be dry enough to prevent microbial growth and subsequent spoilage and cooled before packaging. Package should be small enough so that food can be used soon after the containers are opened. The dried food should be stored in a cool, dry, dark environment.

Summary

Horticulture is the branch of agriculture that deals with the art, science, technology, and business of growing plants, including medicinal plants, fruits, vegetables, nuts, seeds, herbs, sprouts, mushrooms, algae, flowers, seaweeds, and nonfood crops such as grass and ornamental trees and plants.

Horticulture is extremely popular in the United States and world. It produces high-quality food and aesthetics. Horticulture is a multibillion-dollar business. Vegetables and flowers, fruit and nut trees make up a majority of the horticulture enterprise.

Other practices include the nursery, turf, and landscape design and gardening industry.

Home gardens grow a variety of plants for food, fun, and aesthetics. Basic knowledge required for the home gardener is providing the proper location and preparation for the garden as well as selecting the proper variety of plants, time to plant, and proper care for the plants. Some produce from the garden, if not used, can be sold, given away, or preserved for later use.

Special gardens include container gardens and front yard gardens for people with limited space and community gardens for others. Gardeners with little experience can get information online or from the local county extension service, state universities, federal and state agriculture departments, radio, TV, local garden centers, nurseries, and neighbors.

Floriculture is also an international billion-dollar industry. Foliage and flower plants like vegetables are usually grown from seed, but sometimes living plants are taken from cuttings and plant parts and are asexual versus sexual (seed). A floriculture crop is at its highest quality at harvest and must be properly handled to provide a quality product for the consumer.

Growing vegetables, fruits, flowers, and foliage plants in the greenhouse, nursery, and field is a very complex and important industry in the U.S. and world. Marketing and growing quality crops are time-consuming, expensive, but very necessary.

Landscaping and turf is another multibillion-dollar industry to beautify and complement the home gardener. Most homeowners would like a nice lawn around their home as well as a beautiful landscape. Many business locations beautify their surroundings with special lawns, hedges, work, and service areas and special features to improve aesthetics.

Seed production for the horticulture industry and civilization is big business. Plant breeders have provided the world with quality plant and food products. Food preservation and packaging have been discussed.

Chapter 6. Literature Cited

Anonymous (2016). Horticulture - Wikipedia, the free encyclopedia. Retrieved from en.wikipedia.org/wiki/Horticulture. Accessed 5 February 2016.

Shoemaker JS (1952). General Horticulture. J.B. Lippincolt Company, New York. 464 p.

Sperry N. (1982). *Neil Sperry's Complete Guide to Texas Gardening.* Taylor Publishing Company, Dallas, Texas. 497 p.

Boodley JW (1998). *The Commercial Greenhouse.* 2nd Edition. Delmar Publishers. International Thomson Publishing, Inc. Albany, New York. 612 p.

George RAT (1999). *Vegetable Seed Production.* 2nd Edition. CABI Publishing. CAB International Wallingford, OX on UK. 328 p.

Dole JM and Wilkins HF (2005). *Floriculture: Principle and Species.* 2nd Edition. Pearson Prentice Hall, Pearson Educations, Inc. Upper Saddle River, N.J. 1023 p.

Markle GM, Baron JJ, Schneider BA (1998). *Food and Feed Crops of the United States.* 2nd Edition. Meister Publishing Company. Willoughby, OH. 517 p.

AgEdLibrary.com (date unknown). Determining the Importance of the Horticulture Industry. Retrieved from www.senecahs.org/pages/uploaded_files/importance of the E-Unit Determining the Importance of the Horticulture Industry AgEdLibrary.com. Copyright by CAERT, Inc. Accessed 12 February 2016.

Blackwell C (1977). Why Folks Garden and What They Face. Pages 4-14 *in* Hayes J ed. Gardening for Food and Fun. The Yearbook of Agriculture 1977. U.S. Department of Agriculture. Superintendent of Documents, U.S. Government Printing Office. Washington, D.C.

Wilson JW (1977). Where to Garden-Setting Your Sites. Pages 15-23 *in* Hayes J ed. Gardening for Food and Fun. The Yearbook of Agriculture 1977. U.S. Department of Agriculture. Superintendent of Documents, U.S. Government Printing Office, Washington, D.C.

Bartok JW Jr (1977). Garden Tools and Equipment. Pages 24-32 *in* Hayes J ed. Gardening for Food and Fun. The Yearbook of Agriculture 1977. U.S. Department of Agriculture. Superintendent of Documents, U.S. Government Printing Office, Washington, D.C.

Scheel DC (1977). Pest Management Is A Matter of Timing. Pages 71-77 *in* Hayes J ed. Gardening for Food and Fun. The Yearbook of Agriculture 1977. U.S. Department of Agriculture. Superintendent of Documents, U.S. Government Printing Office, Washington, D.C.

Judkins WP (1977). Organic Gardening - Think Mulch. Pages 78-83 *in* Hayes J ed. Gardening for Food and Fun. The Yearbook of Agriculture 1977. U.S. Department of Agriculture. Superintendent of Documents, U.S. Government Printing Office, Washington, D.C.

Combs OB (1977). How Plants Grow - and Let's Hope They Do. Pages 47-53 *in* Hayes J ed. Gardening for Food and Fun. The Yearbook of Agriculture 1977. U.S. Department of Agriculture. Superintendent of Documents, U.S. Government Printing Office, Washington, D.C.

Anonymous (2012) Patent Protection of New Plant Varieties. Retrieved from www.smart-biggar.ca/en/articles_detail. cfin?news_id-686. Accessed 1 March 2016.

Duble RL (1996). *Turfgrasses: Their Management and Use in the Southern Zone.* 2nd Edition. Texas A&M University Press. College station, Texas. 323 p.

Boodley JW (1998). *The Commercial Greenhouse.* 2nd Edition. Delmar Publishers, International Thomson Publishing, Inc. New York, NY. 612 p.

Thompson P (2008). *Seeds, Sex and Civilization: How the Hidden Life of Plants has Shaped Our World.* Thames and Hudson, Ltd, London, England. 272 p.

Zottola EA and Wolf ID (1977). The Whys of Food Preservation. Pages 298-303 *in* Hayes J ed. Gardening for Food and Fun. The Yearbook of Agriculture 1977. U.S. Department of Agriculture. Superintendent of Documents, U.S. Government Printing Office, Washington, D.C.

Kirk DE and Raab CA (1977). Home Drying of Fruits and Vegetables. Pages 356-360 *in* Hayes J ed. Gardening for Food and Fun. The Yearbook of Agriculture 1977. U.S. Department of Agriculture. Superintendent of Documents, U.S. Government Printing Office, Washington, D.C.

FORESTRY AND TREE CROPS

Introduction

The uses of wood in our daily lives and the products of tree crops are almost too numerous to list. However, woody plants have been a dominant part of our civilization for centuries or at least for the recorded history of humans. One is in touch with wood and woody products each day and probably for the foreseeable future. In my college days of the 1950s, we discussed the possible replacement of wood products by synthetic products (plastic) from petroleum but it did not and will not happen.

Major uses of wood include lumber, fuelwood, posts and poles, veneers, plywood and laminated structures, railroad ties, cooperage, shingles, pulp and paper, rayon, plastics. and many other products (Schery 1952).

Woody plants have been on earth an estimated 395 to 400 million years and the chemical composition varies from species to species but is approximately 50% carbon, 42% oxygen, 6% hydrogen, 1% nitrogen, and 1% other elements. Cellulose is a major component at 41-43%, hemicellulose is about 20%, and lignin is the third major component at around 27% in coniferous wood versus 23% in deciduous trees (Anonymous 2016a).

Our food and shelter would be significantly reduced without forest products and tree crops.

Types of Trees

Schery (1952) indicated that lumbermen and foresters recognize two convenient general categories of forests, softwoods and hardwoods based upon the predominant type of trees contained. Botanically softwood is any member of the subphylum *Gymnospermae* while hardwoods belong to the subphylum *Angiospermae*. The terms softwood or hardwood are sometimes almost meaningless, for many softwoods are harder than many hardwoods, although on average hardwoods are collectively more dense and sturdier than softwoods.

Examples of softwood trees (Gymnosperms) include the valuable pines, spruces, and Douglas fir (conifers) and are readily distinguishable because the leaves are usually in the form of needles or scales and the reproductive structures are in cones. Common hardwoods or Angiosperms are such trees as the oaks, elms, maples, and similar species.

Gymnosperms occupy chiefly temperate to subarctic areas with moderate rainfall although minor stands are found in tropical climates, especially at higher altitudes. They become increasingly more common as the climate becomes progressively colder.

In contrast the Angiosperms (hardwoods) are very diverse and are easily distinguished from the Gymnosperms in having broad, deciduous leaves and flowers instead of cones as the reproductive structure. Botanically, Angiosperms are distinguished from Gymnosperms in having the ovule enclosed within the ovary rather than exposed as with cones in Gymnosperms.

Any common hardwood fruit like the peach, walnut, locust pod, or oak acorns (matured ovule) are contained within the outer enclosing tissues that develop from the flower.

World Abundance

Schery (1952) stated that in 1948 the Food and Agriculture Organization of the United Nations estimated that about 36% (2.5 billion acres or a billion hectares) of the forest area of the world is predominantly coniferous. In the United States about 62% of our commercial forest land is in softwoods and about four-fifths of our timber crop is softwood, mainly Douglas fir.

Other estimates indicate that almost half of the total forest area of the world, totaling about 3.6 billion acres (1.5 billion hectares), is of tropical hardwood (Schery 1952). Most of the tropical forest area occurs within Central Africa and the Amazon Valley of South America. The temperate hardwood forests comprising about 16% of the total forest areas of the world occur in east Central North America and scattered through the middle latitudes of Eurasia.

Simmons (1974) estimated that about 4 billion hectares of the world is in forest lands with nearly 1 billion hectares each in North and Central America, South America, and the old USSR. Senguta and Moginis (2005) reported a total of 3.9 billion hectares, or 30% of the land area of the world, covered in natural and plantation forests. Natural forests were at 3.8 billion hectares and plantation forests were about 0.2 billion hectares including Africa, Asia, Europe, North and Central America, Oceania and South America. However, between 2008 and 2012 most countries such as Australia, Brazil, Canada, European Union, Germany, India, Indonesia, Japan, Russia, and the United States all report forest area losses (Anonymous 2016a). The United States Forest Service estimated a net loss of about 2 million hectares (4.9 million acres) between 1997 and 2020. This includes conversion of forest land to other uses as well as afforestation and natural reversion of abandoned crop and pasture land to forest. However, many acres of the United States' forest is stable or increasing, particularly in many northern states.

According to the World Bank in 2011, forested land area of the world is 30.88% and includes natural and planted stands of forests and has been about 31% since 1990 (Anonymous 2016b). Whatever the trends, forests account for 75% of the gross primary productivity of the earth's biosphere and contain 80% of the earth's plant biomass (Anonymous 2016a). Good management will be required to sustain forest productivity in the future.

Brief History

Binkely (2003) indicated that in the last 10,000 years of human existence, man had to obtain sustenance from numerous wild organisms (Chapter 1); but today we get virtually all of our food from about ten domesticated plants and five domesticated animals. So, for agriculture the transition is now complete. Since the production cycle for trees is long-term compared to annual crops, and the inventory of standing timber is largely relative to annual consumption, the transition in our use of forests will take a much longer time than it did for agriculture. This transition still has a few decades to go but helps explain the forest situation today (Binkely 2003).

Timber depletion has driven timber prices to a high level in the world so that the remaining natural forests will be protected or will simply be too costly to log them given the availability of lower-cost plantation-grown trees. Combined with sophisticated wood products technology, the plantation-grown wood can be substituted for the natural forest. However, natural forests will continue to provide many environmental services and products. The recognition of forests in carbon sequestration could result in a high influx of capital into forestry and provide more opportunities for the private sector (Binkely 2003).

Forest Ecosystem Service

Climate Change

Climate change has emerged as an important issue in the world whether one is a protagonist or an antagonist on the issue. Forests are both a source of carbon dioxide when destroyed or degraded and a sink when conserved, managed, planted, and sustained. Forest vegetation and soils hold about 40% of all stored carbon in the terrestrial ecosystems. From the viewpoint of adapting to climate change, conservancy and restoring forests play an important role in redirecting climate-related floods and droughts (Senguta and Moginis 2005).

Water Supply

Forests play an important role in supplying fresh water. Many large cities depend on forests for their water source and much of the world's fresh water supply comes from forests located in the mountains. These forests make up about 28% of the world's forests (Senguta and Moginis 2005).

Biodiversity Conservation

Biodiversity loss is of great concern, especially in tropical areas. Loss of plant and animal diversity of the world species needs immediate attention. Government and donor agencies need to emerge. Implementing programs is getting started but needs significant participation and promotion of ecotourism.

Wood Supply and Forest Plantations

According to Senguta and Moginis (2005), wood and wood products are by far the most efficient and environmentally friendly raw material when compared to steel, aluminum, or concrete. Plantations provide more than 35% of the world's wood supply although they are less than 5% of the global forest resource. Future planted forests will have an increasing role in wood and nonwood products as natural forests decline or are used for conservation and other uses. Many developed and developing countries are encouraging forest plantations to meet their wood needs and carbon sequestration projects.

Breeding Programs

Tree breeding is the application of genetic, reproductive biology, and economic principles to the genetic improvement and management of forest trees in contrast to the selective breeding of livestock, arable crops, and horticultural crops over the last few centuries. The breeding of trees, with the exception of fruit trees, is a relatively new occurrence (Anonymous 2015).

A typical forest tree breeding program starts with selection of superior phenotypes in a natural planted forest. This application of mass selection improves the mean performance of the forest. Offspring is obtained from selected trees and grown in test plantations that act as genetic trials. Based on the tests, the best genotypes among the parents are selected. Selective trees are typically multiplied by either seeds or grafting and seed orchards when the preferred output is improved seed. Alternatively, the best genotypes can be directly propagated by cuttings or in-vitro methods and used directly in clonal plantations. The first system is used with conifers, while the second is typical in some broadleaf trees (poplars, eucalyptus, and others). The objective is to improve yield of wood and wood properties and for pest and disease resistance.

Genetic modification of forest trees species using recombinant DNA techniques has been addressed for virus resistance, insect resistance, lignin content, and herbicide tolerance but there has been no reported commercial production of transgenic trees (Senguta and Moginis 2005).

Further, it is acknowledged that biosafety aspects of genetically-modified trees need careful consideration because of long generation time of trees and the potential for dispersal of pollen and seed over long distances.

New Technology

England (2014) described unmanned aerial vehicles, GIS and mobile mapping technology, and a treemetrics 3D scanner that determines the width, height, and volume of trees as well as wood quality. Unmanned aerial vehicles, quadcopters, hexacopters (drones), fixed wing airplanes and blimps that fly over the designated forest areas provide still images, videos, and a live view through the smartphone.

Potential applications of this technology include forest surveying, site planning, disease, insect and weed detection, fire management and other applications to get maximum value from the forest resource.

John Deere has a ForestSight system to optimize harvesting of forest trees. ForestSight integrates machine data, prognostics, and diagnostic tools for efficient and specific applications (Anonymous 2016c).

The Associated Oregon Loggers, Inc. (Anonymous 2016d) present modern sophisticated forest mechanization, harvesting, and management methods that provide improved safety, productivity, and protection of the forest environment. This includes improved machinery, computer log cutting, data records, fuel use, maintenance of equipment, communication, mapping, roadwork, planting, fertilization, and other benefits and uses.

Traditional wood products are still mostly residential construction. New uses include new paper grades for publishing using high mineral content and for high-performance packaging, sensors to measure wood quality facilitating new uses, technologies that increase wood use in multi-residential and institutional construction, and coatings based on nanotechnology that produces high-grade wood finishes (Anonymous 2016e).

Non-traditional products include biofuels, biochemicals, and biopolymers as a result of new product creation. New processes are being developed to combine wood fiber with plastics to produce new materials of enhanced durability and strength. Lignin is being considered for carbon black, a petroleum product used in making car tires. These and other transformative technologies are in various stages of early research to commercialization from the forest industry (Anonymous 2016e).

Management of Forests

Forest management is a branch of forestry concerned with overall administrative, economic, legal and social aspects as well as scientific and technical aspects such as silviculture, protection, and forest regulation. It also may include aesthetics, recreation, urban values, water use, wilderness, wildlife use, wood products, forest genetic research, grazing of domestic livestock, and other resource values. It may include timber harvesting, planting or replanting the desired species, constructing roads and pathways and prescribed fire or fire prevention (Anonymous 2016f).

Forest management can be very complicated depending on the management required and services obtained. The public has become more involved and services have shifted from timber extraction to more conservation and multi-use of watershed management, plant and animal diversity, and recreation. Forest management varies in intensity to achieve either economic viability (increased timber yields, non-timber forest products, ecosystem services) or ecological criteria (species recovery, fostering of rare species, carbon sequestration) (Anonymous 2016f).

Plantation Forests

A plantation is usually a large piece of land in a tropical or subtropical area where the crop is planted for commercial purposes. Crops grown usually include fast growing trees (often conifers), rubber trees, and various fruits. It may also include cotton, tea, coca, sugarcane, and oil seeds (e.g. oil palms) (Anonymous 2016g).

Commercial plantations are established to produce a high volume of wood in a short time. Plantations are grown by state forestry authorities and/or paper and wood industries and other private owners. Private land owners in the United States grow timber and Christmas trees (Anonymous 2016g).

Plants used in plantations are often genetically altered for improved growth and resistance to pests and disease. Selected individuals grown in seed orchards are used to develop adequate planting material. Wood production is usually higher in a tree plantation than that of natural forests. Fast-growing species may yield 20 to 30 cubic meters or more wood per hectare annually compared to natural stands that may only yield 1 to 3 cubic meters per hectare annually (Anonymous 2016g).

Modern Sawmills

Today computers, conveyors, scanners, lasers, digital cameras, and barcoding systems do most of the work. The saws are flexible bandsaws. Employers check the system when needed to keep it operational (Ranklin 2011). It is a much different method to saw logs than in earlier times.

It is all about having less sawdust and chips and more lumber produced. Computers made modern sawmills possible. Hydraulics do the heavy lifting and operate the machinery and sensing technologies run the drying kiln and detect moisture levels. Metal detectors reveal a hidden nail or metal in the log that could ruin the saw. It is all about increasing efficiency (Ranklin 2011).

Lasers working with the computer can scan a debarked log to yield the most usable lumber. There is a big log sorter that automatically tells the chop saw where to cut. The technology of today combines high-speed digital cameras, computer saws to saw, trim, sort, and grade boards (vision optimization).

The dry finished wood products are ready to use or sell. However, the equipment for the modern sawmill and technology is expensive to install.

The North American Forests

Early pioneers in the Northeast United States began cutting down forest trees in the 1700s and the wood supply seemed unlimited (Walker 1999). Wood-fired boilers and steamers on rivers and locomotives, railroad crossties, wooden water towers, shakes and shingles, furniture, wagons, trestle timber, barrel laths, fencing, housing and a hundred other uses demanded wood. By the mid-1800s the Northeast U.S. had essentially been depleted of wood so wood cutting moved westward to Michigan, Wisconsin, and finally to Minnesota. These northern virgin forests were essentially depleted with the help of disastrous fires by the 1900s.

Historians may attribute the delay in moving southward to the long-established biases toward southern-grown wood and the adequate supply of timber in the North. Others objected to the high resin content and easy splitting quality of southern yellow pines. Also loggers objected to the insects and thousands of small lakes that impeded travel and timber accessibility (Walker 1999). The South's forests were finally open to intensive harvest about 1900 and harvest of forests in the Western U.S. accelerated after World War II.

By the 1960s many westerners objected to logging of the old-growth forests. This public objection continues today. Few of the antagonists understand silviculture procedures in forest ecology and therefore are unwilling to recognize that clearcutting can be an example of good ecology to maintain healthy forest ecosystems (Walker 1999).

Antagonists in many forms of agriculture are usually uninformed and lack understanding of the methods of growth and harvest needed for good animal and plant ecology.

Silviculture is defined as the art of growing trees in managed stands for the production of goods and services, concerned with forest establishment, composition, and growth (Walker 1999). Some three-fourths of the South's pine forest fell into two productivity classes: 40 cords (a cord is a woodpile 8 feet long, 4 feet wide, and 4 feet high) per acre in 20 years and 20 cords per acre in 20 years. Most of this land could produce 80 cords per acre in 20 years or 6 cords per acre per year under intensive management. Loblolly, shortleaf, and slash pines are grown using similar silviculture. Most efficient regeneration calls for clearcutting depending on seeds from the walls of surrounding trees in the 20-acre openings or planting 1-year-old nursery stock seedlings (Walker 1999). Strip clearcutting is also employed with swaths 200 foot wide cut through the forest. Fire prepares seed beds to receive the winged seeds or bulldozers windrow brush and logging slash in order for planters to have easy access to the site.

While management plans may call for selection harvests, these usually are simply thinnings, patch clearcutting, or initial harvests of the shelterwood system.

Intermediate management involves timber stand improvement (TSI), periodic prescribed fire, or herbicide use to control weeds and brush. Thinning is especially effective because it allows the residual trees to respond and grow. Commercial fertilizer sometimes improves flowering and seed production and growth of the trees but may be too costly. Soil sampling is needed to determine nutrients needed before fertilizer application. Weed control is especially important in southern pine forests in order to expedite growth of the conifers. Herbicides, mechanical means, and fire are employed. Insecticides may also be necessary for control of the black turpentine beetle. Other insects, fungi, and pathogens can be problems. Much care and experience is necessary for successful wood and pulp production.

Walker (1999) lists the various forest trees found in the United States by region and state as indicated.

Alaska and Canada	white and black spruce, jack pine, eastern larch and tamarack, northern white-cedar, sitka spruce, western hemlock
Conifer Forests of the Northeast	eastern and northern white pine, red pine, jack pine, eastern hemlock, spruce and fir, eastern larch, eastern redcedar, pitch pine, northern white-cedar

Broadleaf Forests of the North and Mid-Continent	quaking aspen, black cherry, black walnut, American beech, yellow birch, sugar maple, gray birch, white birch, yellow poplar, American elm, rock elm, slippery elm, black green and white ash, cottonwood, sycamore, sweet gum, American chestnut
Pines of the South	loblolly pine, (oaks, hickories, sweet gum are associated), short-leaf pine, slash pine, longleaf pine, pond pine, sand pine, Virginia pine, table-mountain pine, son-deregger pine, pitch pine
Other Conifers of the South	eastern redcedar, southern bald cypress, eastern white pine, pondcypress, Atlantic white-cedar, Fraice fir, balsam fir, red spruce, eastern hemlock
Upland Broadleaf Forests of the South	post oak, cottonwood, eucalyp-tus, black locust, yellow-poplar, live oak, hybrid poplar, paulownia
Broadleaf Forests of Southern Wetlands	bald cypress, water tupelo, willow oak, southern cottonwood, nut-tall oak, cherrybark oak, swamp chestnut oak, sweetgum, live oak, water oak, loblolly pine, white oak, black willow, overcup oaks, shumard oak, American syca-more, yellow-poplar, white oak

Mixed Conifer-Broadleaf Forests of the East	
Northern	red spruce, sugar maple, beech, paper birch, balsam fir, northern red oak, white pine, red maple, chestnut oak, hemlock yellow-birch
Central	yellow-birch, eastern hemlock
Southern	longleaf pine, scrub oak, shortleaf pine, oaks, Virginia pine, loblolly pine, hardwood, slash pine, hardwoods, southern bald cypress, tupelo
Pine Forests of the West	western white pine, ponderosa pine, jeffrey pine, jack pine, lodgepole pine, torrey pine, pinyon pine, sugar pine, whitebark pine, foxtail pine, digger pine
Spruce and Fir Forests of the West	douglas fir, sitka spruce, pacific silver fir, red fir, white fir, subalpine fir, Engelmann fir, blue spruce, aspen, grand fir, noble fir
Other Conifer Forests of the West	western red cedar, western hemlock, mountain hemlock, giant sequoia, redwood, incense-cedar, western larch, alpine larch, Arizona cypress, Port Orford white-cedar, Rocky Mountain juniper, alligator juniper, oneseed juniper, Mexican cedar, western juniper, Alaska cedar

Broadleaf Forests of the West	trembling aspen, tanoak, Pacific madrone, red oak, blue oak, true oaks, Oregon white oak, California coast live oak, canyon live oak, gambel oak, cottonwoods and willows, western rock maple, mesquite, arboreal hardwoods, chaparral, birches and balsam poplar, blue oak, willow oak
Tropical Forests of Hawaii, Southern Florida and Puerto Rico	<u>Hawaii</u> Rainforest Dry forest Arid scrub Alpine stone desert Strand vegetation beaches (coconut) Exotic scrubland <u>South Florida</u> slash pines, mahogany, southern bald cypress, pondcypress, palms and oaks <u>Puerto Rico</u> 3000 kinds of plants palms, bamboos, gnarled ceiba, cyrilla, cadam trees (fence post), mahogany, eucalyptus, Spanish cedar, acacia, mesquite, mango, coffee trees, foxtail pine, slash pine

Trees of the World

The trees of the world are too numerous to list in this chapter, but there are three main types of trees that are popularly recognized in terms of leaf character (Anonymous 1981). There are conifers with needle-like and scale leaves, the broadleaves which have flattened simple or compound leaves, and the monocotyledon trees like the palms in which they are mainly fan-like or fern-like in shape (Anonymous 1981).

In today's world the broadleaves are the dominant natural tree-form throughout the world except in extreme cooler climates where conifers still prevail. The palms are mainly tropical with only a few of the 2800 species occurring in the subtropics and temperate regions.

We have mentioned the pines, firs, spruce, birches, maples, oaks, aspen, hemlock, cedars, larch, junipers, willows, cottonwood and poplars of North America grown for wood but have not mentioned those grown for shade, beauty, and food crops except in Chapter 6.

In 1950 Smith (1953) highly recommended tree crops as a permanent agriculture like chestnut orchards for annual nut or fruit crops for humans and livestock. This also includes walnut, pecan, hickory, persimmons, mulberries, acorns from oaks, carob, kiawe, honey locust, mesquite, figs, olives, or fruit trees. Annual crops can sometimes be grown with tree crops or the tree crop can be grown on soils and rocky areas not suitable for annual crops.

Some mention of fruit and nut crops were given in Chapter 6. They include apples, pears, plums, apricots, peaches, avocados, cherries, grapes, olives, nectarines, and figs. Nut crops include almonds, pistachios, walnuts, pecans, filberts, and hazelnuts. Citrus crops include grapefruits, lemons, limes, oranges, tangelos, and tangerines.

Tree Culture

The time to plant fruit and nut crops can be mid-September through November in the southern United States and fall in the northern United States. They are usually sold bare rooted and fall planting gives them time to establish new root growth before spring. Proper spacing depends upon the kind of plant; but fruit plants require adequate spacing, which is usually 25 to 30 feet between fruit trees. Pecan trees require 35 to 45 feet between plants. Consult your local county agent or garden center for proper care of the trees in planting, pruning, fertilizing, watering, controlling pests, and harvesting.

Shade Trees

Many woody plants can serve as beauty and shade trees including some of the fruit and nut trees. About any conifer, broad-leaved or monocotyledon tree, will be suitable depending on preference and climate adaptation. Shade trees for the United States are listed as follows:

Northeast
maples (sugar, Norway)
elms (English, Chinese, Siberian)
oaks (pin, northern red oak, scarlet, white, willow, Shumard or Texas)
ginkgo
tulip tree or yellow poplar
sweetgum
American linden or basswood
American sycamore or the planetree
London planetree
American yellowwood
European beech

thornless common honeylocust
common hackberry
eastern hemlock, red pine, white fir
Nikko fir, Oriental spruce, Colorado,
native white spruce, blue spruce and many others
(Waterman, Swingle, and Moses 1949)

Southeast
Southern magnolia, the camphor tree
willow oak, red maple, flowering dogwood
sweetgum, American holly
American beech, eastern redbud
water oak, winged elm, American elm
(Lindgren, True and Toole 1949)

The Plains
green ash, American elm, Siberian elm
common hackberry, bur oak
ailanthus or tree-of-Heaven, boxelder
northern catalpa, Kentucky coffeetree
Chinese elm, sugar hackberry
common honey locust, silver maple
Russian-olive, American sycamore
London planetree, weeping willow
eastern black walnut
eastern redcedar
ponderosa pine, Douglas fir
white fir, Austrian pine
spruces
(Wright and Bretz 1949)

The Rockies
tree-of-Heaven, green ash, velvet ash
white ash, boxelder, northern catalpa
American elm, Siberian elm
common hackberry, thornless honey locust
linden, black locust, Norway maple
Russian mulberry, bur oak
London planetree, lanceleaf poplar
Lombardy poplar, plains poplar
Russian-olive, tomarisk
Arizona cypress, eucalyptus
Rocky Mountain juniper
aleppo pine, Austrian pine
canary pine, ponderosa pine
Colorado pinyon pine, scotch pine
Colorado spruce, Engelmann spruce
and others
(Gill 1949)

California
California live oak, camphor tree
red ironbark, cape chestnut
California pepper tree, ginkgo
Norway maple, London planetree
sweetgum, velvet ash, Carolina poplar
pin oak, southern red oak, California black walnut
Chinese pistache
panicled golden rain tree
white mulberry, canary pine, Coulter pine
Lawson cypress, California incense-cedar
Deodar cedar
(Wagener 1949)

North Pacific
common hackberry, sweetgum
American yellowwood, yellow-poplar
northern red oak, Oregon white oak
pin oak, bigleaf maple
American elm, atlas cedar
Lawless cypress
Himalayan pine, Douglas fir
California incense-cedar
(Childs 1949)

Fruit and Nut Crops
As indicated, the growing of such crops as apples, plums, pears, apricots, peaches, avocados, pecans, walnuts and citrus crops require extensive care and are long-term. It will require several years after planting before fruit and nut trees will produce and need to be protected from harmful weather, insects, pathogens, and weed competition. In an orchard operation, some trees may die early and need to be replaced. A rotation system to bring in new plantings in a large operation would help sustain production. These are intensively managed crops with high risk and need to be grown in fertile soils with an adequate water supply. They add millions of dollars to our economy and provide great nourishment and variety to our food supply. Trees need to be pruned and dead ones removed as appropriate. A warm and long growing season is required for citrus and trees can be damaged by freezing air temperatures. Cold-storage plants and refrigeration have played an important role in the fruit industry, particularly for fruits that can be stored for long periods such as apples and citrus fruits. Harvests are labor-intensive and timing is critical in removing the crop from the field to preserve and present a wholesome, attractive product (Ebeling 1979).

Summary

Wood has played a dominant role in human civilization for weapons, treasures, shelter, fire, and well-being and will continue to do so for the foreseeable future. Woody plants have been on earth an estimated 400 million years and the major components are cellulose, hemicellulose, and lignin.

Forest trees can be divided into softwoods like the conifers or hardwoods like the oaks, elms, and maples. About 31% of the world is forested lands and includes natural and planted stands but represent about 80% of earth's plant biomass.

Forest vegetation and soils hold about 40% of all stored carbon in the terrestrial ecosystem, play an important role in the water supply, and provide biodiversity for the plant and animal kingdom.

Wood products are by far the most efficient and environmentally friendly material compared to steel, aluminum, or concrete. Plantations provide more than 35% of the world's wood supply although they are less than 5% of the global forest resource. Plantations are much more productive than natural stands of forests.

In the breeding programs, superior phenotypes are selected from natural planted forests. Based on tests, the best genotypes among the parents are selected. Selected trees are multiplied by seeds or grafting. Alternately, the best genotypes can be directly propagated by cuttings or *in-vitro* methods and used directly in clonal plantations. The first system is used with the conifers while the second is typical with the broadleaf trees. The objective is to improve yield and quality of wood products.

New technology consists of unmanned aerial vehicles, GIS, and mobile mapping systems, scanners, and smartphones to determine the width, height, volume, and quality of the trees and wood. Potential application is for forest surveying, site planning, disease, insect and weed detection, fire management, and other applications to get maximum value from the forest resource. Sophisticated equipment is available for forest management, mechanization, harvesting, safety, log sawing, recording data, planting new crops, roadwork, and other management needs.

New products include biofuels, biochemicals, and biopolymers as a result of new product creation as well as new processes for enhanced durability and strength of wood products.

A list of various forest trees found in the United States as well as shade trees are given. A discussion of fruit and nut crops from trees is also given.

Chapter 7. Literature Cited

Schery RW (1952). *Plants for Man.* Prentice-Hall, Inc. Englewood Cliffs, NJ. 564 p.

Smith JR (1953). *Tree Crops: A Permanent Agriculture.* The Devin-Adair Company, New York, NY. 408p.

Simmons IG (1974). *The Ecology of Natural Resources.* John Wiley and Sons, New York, NY. 424   .

Anonymous (2016a). Wood. Wikipedia, the Free Encyclopedia. Retrieved from en.wikipedia.org/wiki/Wood. Accessed 24 March 2016.

Anonymous (2016b). Forest Area (% of Land Area) in World. Retrieved from www.tradingeconomics.com/world/forest-area... land-area-wb... Accessed 26 March 2016.

Binkley CS (2003). Forestry in the Long Sweep of History. Pages 1-7 in Teeter L, Cashore B, Zhang D eds. Forest Policy for Private Forestry: Global and Regional Challenges. CAB International, New York, NY.

Senguita S and Maginis S (2005). Forests and Development: Where Do We Stand? Pages 19-58 in Sayer J ed. Forestry and Development. Earthscan, London. UK and Sterling, VA, U.S.A.

Anonymous (2015). Tree Breeding. Retrieved from en.wikipedia.org/wiki/Tree_breeding. Accessed 4 April 2016.

England J (2014). New Technology in the Forest. South of Scotland Cluster Meeting. Retrieved from www. forestryscotland.com/media.282670/jcpresentationv2.pdf. Accessed 5 April 2016.

Anonymous (2016c). Forestry Technology Solutions. John Deere. Retrieved from www.deere.com/.../forestry-technology-solutions.page. Accessed 5 April 2016.

Anonymous (2016d). Technology Improvement in Logging. Associated Oregon Loggers, Inc., 2015. Madrona Avenue, Salem, OR. Retrieved from www.oregonloggers.org/Forest_Logging_Technology.aspx. Accessed 5 April 2016.

Anonymous (2016e). Transformative Technologies Natural Resources Canada. Government of Canada. Retrieved from www.nrcan.gc.ca/forests/industry/products-applications/13343v. Accessed 7 April 2016.

Anonymous (2016f). Forest Management - Wikipedia, the Free Encyclopedia. Retrieved from en.wikipedia.org/wiki/Forest_Management. Accessed 7 April 2016.

Anonymous (2016g). Plantations. Wikipedia, the Free Encyclopedia. Retrieved from en.wikipedia.org.wiki/Plantation. Accessed 7 April 2016.

Ranklin J (2011). The Modern Sawmill: A High Tech Marvel. Forests for Maine's Future. Retrieved from www.forestsformainesfuture.org/.../the-modern-sawmill-a-high-tech-mar... Accessed 8 April 2016.

Walker LC (1999). *The North American Forests: Geography, Ecology and Silviculture.* CRC Press, LLC. Boca Raton, Florida. 398 p.

Anonymous (1981). Trees of Every Kind. Pages 54-260 in Hora B. ed. The Oxford Encyclopedia of the Trees of the World. Oxford University Press. Oxford. Walton Street, Oxford, $OX_2$6DP.

Lindgren RM, True RP, Toole ER (1949). Shade Trees for the Southeast. Pages 60-65 in Stefferud A. ed. Trees. The Yearbook of Agriculture 1949. United States Department of Agriculture. U.S. Government Printings Office, Washington D.C.

Waterman AM, Swingle RU, Moses CS (1949). Shade Trees for the Northwest. Pages 48-60 in Stefferud A. ed. Trees. The Yearbook of Agriculture 1949. United States Department of Agriculture. U.S. Government Printings Office, Washington D.C.

Wright E, Bretz TW (1949). Shade Trees for the Plains. Pages 65-72 in Stefferud A. ed. Trees. The Yearbook of Agriculture 1949. United States Department of Agriculture. U.S. Government Printings Office, Washington D.C.

Gill LS (1949). Shade Trees for the Rockies. Pages 72-76 in Stefferud A. ed. Trees. The Yearbook of Agriculture 1949. United States Department of Agriculture. U.S. Government Printings Office, Washington D.C.

Wagner WW (1949). Shade Trees for California. Pages 77-82 in Stefferud A. ed. Trees. The Yearbook of Agriculture 1949. United States Department of Agriculture. U.S. Government Printings Office, Washington D.C.

Childs TW (1949) Shade Trees for the North Pacific Areas. Pages 82-90 in Stefferud A. ed. Trees. The Yearbook of Agriculture 1949. United States Department of Agriculture. U.S. Government Printings Office, Washington D.C.

GRAZING LANDS AND PASTURES

Introduction

Lubowske et al. (2006) indicated the land area of the United States was nearly 2.3 billion acres of which grassland pasture and rangeland are 587 million acres (25.9%). Grazing lands are almost equal to forest land at 651 million acres (28.8%) as discussed in Chapter 7. Cropland in the United States occupies 442 million acres (19.5%). Special uses like parks and wildlife occupy 297 million acres (13.1%). Urban lands are 60 million acres (2.6%) and miscellaneous of 228 million acres or 10.1% make up the total.

Grazing of domestic livestock and wildlife also graze and use forests, parks, and wildlife areas so probably at least 50% of the total U.S. land area is used by livestock and wildlife in some way.

In 1948 Wooten and Barnes indicated the grasslands, hay lands, and forested rangelands of the entire United States cover more than a billion acres or nearly 60% of the total land area. This apparently did not include Alaska. However, since 1948 much of the land has been converted into other uses. If we examine the area suggested by Lubowske et al. (2006) including Alaska, grazelands may still approach one billion acres if forest land, special use land, and miscellaneous acres are included. However, Lark et al. (2015) reported several million acres of grassland and other nonagricultural lands converted to cultivation of corn and soybeans following the biofuels boom of the 2000s. Cropland expansion occurred most rapidly on land that is less suitable for cultivation, raising concerns about adverse environmental and economic costs of conversion.

Grazing Land Definitions

All areas of the world that are not barren deserts, farmed, or covered by bare soil, rock, ice, or concrete can be classified as rangelands (Holechek et al. 2002). Therefore, rangelands include deserts and forests and all-natural grasslands. Holechek et al. (2002) therefore defined rangeland as uncultivated land that cannot only provide life's necessities for grazing and browsing animals that may also include a host of activities and products, but grazing and browsing animals are the main focus.

Pasture lands are different from rangeland by the fact that periodic cultivation is used to maintain introduced non-native forage species and agronomic inputs such as irrigation and fertilization.

Products of Grazing Lands

The main focus of grazing lands is the production of wool (sheep) and red meat (cattle, goats, sheep). It provides space for tremendous diversity of wildlife and wildlife numbers. They provide huge water reserves and yield for human and animal use. Rangelands provide plant diversity, native plants and ornamentals, and protection of endangered species. It provides recreation such as hunting, fishing, wildlife viewing, picnics and beautiful scenery, wood products and minerals and human habitation. Renewable and nonrenewable fossil fuels (oil, natural gas, coal) are produced and it can serve as a natural sink for CO_2 sequestration and environmental maintenance. Renewable energy sources include wind, solar, and biomass and are in their early stages of development. However, rangeland will provide space for these and other technologies yet to be developed (Holechek et al. 2002; 2015: Holechek and Sawalhah 2014).

Problems of Grazing Lands

Although rangelands and pastures provide many necessary products for human and animal well-being, as with any resource there are certain problems to be dealt with. First, uncertain rainfall and adverse weather may limit the primary products of production and habitation. Periods of drought and overgrazing may cause desertification. The biological productivity may decline while the prevailing climatic conditions are thought to remain constant. Human activities such as overgrazing, burning, woodcutting, and temporary cultivation are implicated; and application of sound range management practices have considerable potential to reduce or reverse desertification in many parts of the world.

Since the human population is rapidly expanding, reduction of open space and land area is in decline and less space will be available for the products of grazing lands.

Because of sometimes poor management, weeds, brush, and poisonous plants become problems and significantly reduce forage production and cause grazing losses. Predators (coyotes, wolves, etc.) may cause significant livestock loss. Insects may cause loss of animal vigor and weight. Grazing lands may harbor insects and rodents that are destructive to grazing lands and destructive to nearby cropland.

Types of Rangelands

Grasslands, desert shrubland, savanna woodlands, forests, and tundra are the basic rangeland types of the world (Holechek et al. 2002).

Grasslands

Grasslands are the most productive rangelands in the world when forage production is considered. Grasslands are typically free of woody plants and occur from sea level to 5000 meters in elevation. They are dominated by plants in the Gramineae family (grasses). Forbs (broadleaves) are also important in many grasslands and add important nutrients to the diets of wildlife and domestic livestock. Sedges, rushes, and shrubs (woody) may also be a minor component of grasslands. Grasslands generally occur in areas receiving 250 to 900 mm of annual precipitation. Soils are usually deep (over 2 meters), loamy in texture, high in organic matter, and fertile (Mollisols). These factors make them highly suitable for cultivation, sometimes to the detriment of permanent grassland (Holechek et al. 2002).

Desert Shrubland

Desert shrublands are the driest of the world's rangelands and cover a very large area. Woody plants less than 3 meters high occur with sparse herbaceous understory cover. Desert shrublands have been degraded by heavy grazing in many parts of the world creating desertification and slow recovery. They receive less than 250 mm of annual precipitation.

Savanna Woodlands

Savanna woodlands are dominated by low-growing trees less than 12 meters tall.

Overgrazing causes loss of grasses and increased density of trees and shrubs. They can be transition zones between grassland and forests.

Forests

Forests are different from savanna woodlands by having trees taller than 12 meters in height and denser in space (less than 10 meters apart). Many forests are managed for timber because they are too dense for grazing but may produce forage for wildlife and livestock when thinned by harvest (logging) or fire. Forests generally occur in high-rainfall areas of over 500 mm annually.

Tundra

A tundra is a treeless plain in arctic or high-elevation regions. It covers about 5% of the world's surface. Large areas occur in North America, Greenland, Northern Europe, and Northern Asia. Low-growing, tufted perennial plants and lichens dominate. The genus *Salix* shrubs are the main woody plants. The 7-month permafrost restricts tree growth and precipitation is below 250-500 mm annually on much of the area. Strong winds also restrict plant growth.

Rangelands of the United States

Holechek et al. (2002) indicated there are 15 basic rangeland types in the United States and they are classified as grasslands, desert shrublands, woodlands, or tundra. Grasslands include the tallgrass prairie, southern mixed prairie, northern mixed prairie, shortgrass prairie and palouse prairie and the California annual grassland. Woodlands include the pinyon-juniper woodland, mountain shrubland, western coniferous forest, southern pine forest, eastern deciduous forest, and the oak woodland. Desert shrublands include the hot and cold deserts. The last is the Alpine tundra.

Tallgrass Prairie

The tallgrass prairie is in the central U.S. of the mixed and short-grass prairies and west of the deciduous forests. The climate is subhumid, mesic, and temperate. Precipitation is nearly 1000 mm annually and most occurs during summer. The soils are primarily Mollisols (see Chapter 2). They are deep and fertile. Corn and wheat can be grown in these soils. The American bison once roamed this area. Four grasses typify the tallgrass prairie including little bluestem (*Schizachyrium scoparium*), big bluestem (*Andropogon scoparium*), yellow Indiangrass (*Sorghastrum nutans*) and switchgrass (*Panicum virgatum*). Some other grasses, broadleaf plants, shrubs and trees can occur (Holechek et al. 2002).

Southern Mixed Prairie

The southern mixed prairie extends from eastern New Mexico to eastern Texas and from southern Oklahoma to northern Mexico. Precipitation varies from 300 mm in eastern New Mexico to 700 mm annually in central Texas. Soils are primarily Mollisols, Entisols and Aridisols. The southern mixed prairie consists of true mixed prairie, desert prairie, high plains bluestem, and oak savannah. Mesquite and other herbaceous and woody plants have become weed problems. Game animals are being used to generate income from hunting fees in addition to raising livestock. It is the most important prairie for livestock production of the western ranges (Holechek et al. 2002).

Northern Mixed Prairie

The northern mixed prairie includes the western half of North and South Dakota to the eastern two-thirds of Montana, the northeastern one-fourth of Wyoming, the southeastern part of Alberta, Canada and southern Saskatchewan, Canada. The climate has long, severe winters and warm summers. Precipitation ranges from 300 to 650 mm annually and most occurs during spring and summer. Major soil is Mollisol. It contains short, mid, and tall grasses as well as cool and warm-season grasses and supports both wildlife and livestock. Because of the severe winters, winter supplemental feeding is required. Cool-season grasses such as bluebunch wheatgrass (*Pseudoroegneria spicata*) and various bluegrasses (*Poa spp.*) provide early spring feed. Green needlegrass (*Stipa viridula*), needle and thread (*Stipa comata*), western wheatgrass and various forbs provide late spring feed and little bluestem, blue grama and sideoats grama provide quality summer and fall forage. It is the second most important western range type for livestock production (Holechek et al. 2002).

Shortgrass Prairie

The shortgrass prairie extends from northern New Mexico to northern Wyoming. It is the third most important western rangeland type for livestock production. The area has cool winters and warm summers. Annual precipitation is between 300 to 500 mm with a majority occurring during the growing season.

Soils are mostly Mollisols but soils of Entisol and clay soils of Vertisol occur in the area. Mid grasses such as little bluestem occupy the sandy soils, while western wheatgrass occupies clay soils. Medium textured soils support primarily blue grama and buffalo grass which is 70 to 90% of the shortgrass prairie vegetation. Western wheatgrass, winterfat (shrub) and scarlet globemallow (forb) also occur. Weeds are cactus, snakeweed, Russian thistle, and fringed sagewort. (Holechek et al. 2002).

California Annual Grasslands

The California grassland is found primarily in central California west of the Sierra Nevada and in some parts of Oregon west of the Cascade Mountains. The climate is Mediterranean and is characterized by mild, wet winters and long, hot, dry summers. Rainfall varies from 200 mm in the southern region to almost 1000 mm in some areas near the coast. Most precipitation occurs during winter. Soils vary from Mollisols and Aridisols (desert) to volcanic ash (Inceptisols) in Oregon. The original vegetation was mostly cool-season bunchgrasses in the genus *Stipa*. Today nearly 400 species of introduced annual plants from many places are found in California. Because of this, inadequate forage quantity occurs in fall and winter with inadequate forage quality in summer. Feed supplementation is required during these periods. Brush invasion is also a problem on these rangelands (Holechek et al. 2002).

Palouse Prairie

A large percentage of the Palouse Prairie, also known as the northwest bunchgrass prairie, has been converted into farmland. It occurs in eastern Washington, northcentral and northeastern Oregon, and western Idaho. It evolved with little grazing pressure from large herbivores (no bison). Native grasses were bluebunch wheatgrass and Idaho fescue and were rapidly overused because of low grazing resistance. Rainfall is typically 300 to 640 mm annually with most falling during winter. Winters are relatively mild and summer temperatures seldom exceed 35°C. Soils are deep loessal dunes and mostly in the order Mollisol and are very productive (Holechek et al. 2002).

Hot Desert

The hot desert is one of the largest western range types but low in livestock production because of limited precipitation and high air temperatures. It is found in southern California, Arizona, New Mexico, southwestern Texas, Nevada, and northern Mexico at elevations from 925 to 1900 meters. Precipitation varies from 130 to 500 mm and occurs in winter and summer with most in July, August, and September. The Mojave, Sonoran, and Chihuahuan deserts occur in this type. The areas support a mixture of shrubs and grasses. Mesquite and creosote bush are common. Most plants are warm-season. Grazing is not practical on much of the Mojave and Sonoran deserts. The major soil order is Aridisol (Holechek et al. 2002).

Cold Desert

The cold desert or Great Basin is comprised of the sagebrush grassland and the salt desert. There are about 39 million hectares of sagebrush range with about 65% controlled by the Bureau of Land Management. The remaining 35% is privately owned. It occurs in Oregon, Idaho, Nevada, Utah, Montana, and Wyoming. Precipitation is between 200 and 500 mm annually. Temperatures vary from -37°C in winter to 38°C in summer. The high levels of volatile oil in sagebrush make them unpalatable to livestock but are used by mule deer and pronghorn big game animals. Several different types of sagebrush (big, black, low) can be found in this area. Important grasses include bluebunch wheatgrass, bottlebrush, squirreltail, Idaho fescue, western wheatgrass, Indian ricegrass, needle-and-thread, and great basin wildrye. Crested wheatgrass has been successfully established where big sagebrush has been controlled. Soils are mostly Aridisols (Holechek et al. 2002).

The salt desert shrubland occurs primarily in Utah and Nevada and is not very productive for livestock because of low annual precipitation (80 to 250 mm). It is used primarily as winter range for sheep. Much of the area has been invaded by halogeton, an exotic weed toxic to sheep. Tall wheatgrass has been established on some of the wetter sites after brush control. Most of the area is controlled by the Bureau of Land Management. Soils are mostly Aridisols (Holechek et al. 2002).

Piñon-Juniper Woodland

The piñon-juniper range type occurs from Washington state to north of Mexico City east of the Cascades and Sierra Mountains with most in Utah, Colorado, New Mexico, and Arizona. It has low precipitation, hot summers, high wind, low relative humidity, high evaporation rates, and intense sunlight. Soils are poorly developed and are mostly Entisols and Aridisols. Due to overgrazing, lack of fires, and spread of seed by birds and mammals, trees have encroached. Wood from the piñon and juniper trees has become more valuable than the depleted range. Tree removal is necessary to increase livestock and wildlife forage (Holechek et al. 2002).

Mountain Browse

The mountain browse range type occurs primarily in the Rocky Mountains and Sierra-Cascade Mountains of the western U.S. It's a narrow strip between the grasslands described above and occurs above the piñon-juniper type in the intermountain area of Colorado, Utah, Oregon, and Idaho. Precipitation averages about 460 to 500 mm and air temperatures range from 35°C in summer to -34°C in winter. The growing season is 100 to 120 days. Shrubs dominate, including chokecherry and buckbrush.

It is an important big game winter range. Soils are mostly Entisols and Inceptisols (Holechek et al. 2002).

Oak Woodland

This type is dominated by oak species. The post oak savanna of Texas has from 660 to 800 mm of precipitation annually. The chaparral ranges of southern Arizona and New Mexico average about 360 mm precipitation per year and the Gambel oak ranges in central and southern Rocky Mountains average from 380 to 500 mm of precipitation per year. The California chaparral ranges receive from 650 to 1000 mm precipitation annually. Oaks are sensitive to winter cold and are not found much farther north than north central Oregon. Oaks are important to wildlife but have limited value for livestock (Holechek et al. 2002).

Western Coniferous Forest

The western coniferous forest is mostly ponderosa pine and Douglas fir on about 17 million hectares in the interior western U.S. Precipitation ranges from 450 to 650 mm annually with most as snow in the north and rain in the south. These areas are grazed by livestock but multiple use of recreation, watershed, wildlife, and scenic values are becoming more important than livestock production. Soils are Entisols with Inceptisols on ridges and benches and Mollisols on gentle topography in the ponderosa pine ranges and Alfisols, Entisols, and Inceptisols in the Douglas fir-aspen zone above the ponderosa pine zone (Holechek et al. 2002).

Alpine Tundra

The alpine tundra is the highest range type in altitude and occurs above the spruce-fir type. Alpine ecosystems occupy mountain areas above the timberline and have short, cool growing seasons and long, cold winters. The low-growing vegetation (20 cm or less in height) is characterized by perennial, herbaceous, shrubby vascular plants, extensive mats of cryptograms (mosses and lichens) and the complete absence of trees due to permafrost. Alpine ecosystems are found mainly in Alaska, Colorado, Washington, Montana, and California. They receive about 1000 to 1500 mm precipitation annually, mostly as snow. Members of the bluegrass and sedge families occur in alpine areas. The alpine areas are a source of water for the western U.S. Soils are Entisols to boggy Histosols (Holechek et al. 2002).

Southern Pine Forest

The southern pine forest is one of the largest and most important range types in the U.S. for livestock production. Precipitation averages 1250 mm or more per year and is distributed fairly evenly during the year after the dry fall. Climax vegetation is oak-hickory. Most of the pine is a longleaf, shortleaf, and loblolly. Most is grown for lumber and pulp. Prescribed burning in winter is practiced to eliminate mulch and old growth. Livestock and wildlife utilize these areas. Soil is Ultisol (Holechek et al. 2002).

Eastern Deciduous Forest

The Eastern deciduous forest is important to livestock. It occurs in large areas of Missouri, Indiana, Ohio, Kentucky, Virginia, and Wisconsin. Annual precipitation is between 800 to 2000 mm and is uniformly distributed throughout the year. The growing season is 120 to 140 days. Soils are primarily Alfisol. Most of the forest area has been modified by farming, logging, and industrialization. Maples, birches, oaks, hickories, beeches, and basswoods predominate. Grasses are bluestem, fescue, bluegrass, ryegrass, and orchardgrass. Rotational grazing is preferred. Legumes, particularly clover, are often grown with pasture grasses to improve livestock nutrition and soil fertility (Holechek et al. 2002).

Management of Rangelands

If you are a rancher or land manager in one of the rangeland areas just described by Holechek et al. (2002), then you should know the potential and limitations of the land for livestock and other uses. Such knowledge is gained by personal experience, reading the literature, consulting with the local and state extension service or professional range managers, and visiting with neighboring ranchers. Special programs for land managers are sometimes produced by state university extension and industry personnel regarding carrying capacity or stocking rates, weed and brush control, grazing systems and management, dealing with drought and adverse weather, predators of livestock, forage restoration and maintenance, and a host of other important issues.

Physical Characteristics

Climate, soil, and topography determine the type of vegetation and ultimately the productivity of a given area. These factors determine the types and numbers of livestock and other uses that can be made of the area in question.

Precipitation

Precipitation is extremely important and forage production is directly influenced by it. Most rangelands of the world receive less than 500 mm rainfall annually. Drought on rangeland can be prolonged and significantly reduce stocking rates of livestock and forage production of a given area. Wind can also substantially reduce precipitation efficiency by increasing water loss from soils and plants. Temperature influences forage production if too high or too low during critical growth periods of forage growth. Temperatures above average usually occurred during periods of drought, adversely adding to the problem. Low temperatures at high elevation limit the number of frost-free days and limit the time plants can complete their life cycle (see Chapter 3. Water).

Soils

Soils were discussed in detail in Chapter 2. Knowledge of soils is essential for the rangeland manager. Most soils in the western United States are slightly basic because they receive low precipitation and are derived from calcareous parent material. Those in the eastern United States are slightly acidic because of leaching of soil minerals. Those of the southeastern United States have been most heavily leached. As indicated in Chapter 2, texture, structure, depth, pH, organic matter, and fertility are important in their productivity on rangelands. However, much of the rangeland soil is high in pH, has poor texture and structure, and low fertility. Even with these limitations, rangelands can be productive taking care not to overgraze and misuse them.

The soil classification system for the United States as explained in Chapter 2 includes eleven Orders and they are Enisols, Inceptisols, Andesols, Aridisols,Vertisols, Mollisols, Alfisols, Spodosols, Ultisols, Oxisols, and Histosols. A majority of U.S. rangelands are Mollisols, Entisols, and Aridisols as indicated in the rangeland types described above by Holechek et al. (2002). Mollisols are the natural grassland or prairie soils that are typically deep with high organic matter and have a moderate profile development. Entisols, however, are found in the Rocky Mountains; they are young in development and lack horizon development.

Aridisols are soils found in the Southwest with little profile development and are dry for extended periods (desert climate).

Other soils of rangeland include Inceptisols that are slightly more advanced than Entisols. They are formed by volcanic ash and are found in the Pacific Northwest. Vertisols are found in the south central U.S. and have clays that swell when wet and shrink and crack when dry. Ultisols are highly leached soils primarily in the Southwest and support forest or savannah woodland. The A horizon is leached and the B horizon has an accumulation of clay. Alfisols are similar to Mollisols but have higher organic matter. These soils are associated with the Eastern deciduous forest.

Topography

Wide differences occur in climate and vegetation with location and elevation. Elevation can affect forage growth dramatically in the spring if snow or low temperatures reduce plant growth and affect livestock use. In the fall, animals should be removed on high elevation ranges because of the onset of winter. Vegetation differences also occur on slope orientation, especially from north- versus south-facing slopes. In spring grazing animals may find more forage on warmer south and west-facing slopes and prefer them during cold months because of the warmer temperatures. Degree of slope is also important to range animals. As slope increases, vegetation productivity declines, as well as animal use, because of difficulty in maneuvering the terrain.

Vegetation Monitoring and Assessment

Each rancher or land manager needs to take inventory and assess the vegetation, whether private or government land, to determine its condition and future use. Traditional evaluation or first step in an inventory program involves development of a vegetation map. This can be done by aerial photography and remote sensing techniques. There are commercial laboratories that can help with this analysis. Help from the Natural Resources

Conservation Services (NRCS) is also available, as well as the state extension service professional rangeland consultants and state university personnel.

Actual sampling or estimating of the standing crop can be used to determine grazing capacity, range condition, watershed evaluation, and wildlife use and quality. Determination of grazing capacity can be taken from grazing surveys provided by federal agencies like the Forest Service or Bureau of Land Management. However, climatic fluctuations, wildlife uses, livestock grazing method and intensity, and changes in land managers may alter grazing capacity evaluation.

Historically the amount of climax vegetation found on rangeland sites has been used as an important measure of range condition. However, soil stability and suitability for use and ecological position should be considered for a satisfactory range condition rating (Holechek et al. 2002). The kind of vegetation found and its seasonal productivity will also determine its grazing capacity and use.

Other Management Needs

Stocking Rate

The terms AUM, stocking rate, and carrying capacity are terms widely used in range management. <u>AUM or Animal Unit Month</u> is the amount of feed required to sustain a 1000-pound cow and her calf (up to six months of age) for one month (roughly 800 pounds of oven dry forage). <u>Carrying capacity </u>is the maximum number of individuals of a given species that a site can support over a certain period of time without causing deterioration of the site. (How many animals the forage base is capable of supporting - subject to change due to growing conditions). <u>Stocking</u>

<u>rate</u> is the number of animals per unit area in a given period. (How long your animals are out on your pasture).

Calculations of these terms are available and will not be repeated here (Holechek et al. 2002; Hancock 2006), but proper stocking rate is extremely important in range management. Light grazing intensities are necessary to sustain arid environments. Areas with a long grazing season with high amounts of precipitation on flat terrain with deep soils can tolerate heavier grazing intensity than arid sites. Conservative stocking rates that give good financial returns on a long-term basis and conserve or improve range condition are desired.

Grazing Method

To understand different types of grazing methods is to understand continuous and rotational grazing. Rotational grazing implies only one section of pasture is grazed at a time while the remainder of the pasture is allowed to grow. Rotational grazing allows the plant to remain healthy by renewing energy reserves, rebuilding plant vigor, and giving long-term maximum production. Animals having unrestricted and uninterrupted access throughout the grazing season are continuous grazers.

Research has shown that specialized grazing systems give either modest (10 to 30%) increases in stocking rates or no increase in carrying capacity over season-long or continuous systems (Holechek et al. 2002).

Weed Control and Revegetation

Unpalatable or poisonous plants occur on at least one-half of the rangelands of the United States. Much of the area is covered in a mix of herbaceous and woody plants. Many of the weeds on western rangeland are alien weed species that have no natural enemies (insects or diseases) in the United States. The reason for these problems may be competition (selection of desirable species by livestock and rejection of weeds), reduction of fire, transport of seed by grazing animals and birds, drought, and other disturbances.

Once weeds become dense, some form of control may be necessary in order to have a profitable enterprise. Weed and brush control is accomplished by chemical, mechanical, prescribed fire, or biological means. Chemicals are usually expensive so they are usually used on the more productive sites. Mechanical methods of bulldozing, disking, shredding, roller chopping, chaining, root plowing, skid-steer loading, or grubbing are also expensive and are usually done to control woody plants.

Some biological methods such as insects have been successful but are generally limited to St. John's Wort in the United States and prickly pear cactus in Australia. Goats have been successful in brush suppression. Prescribed fire is a very useful tool in regeneration of rangelands and for successful weed control. It requires experienced managers and technicians for safety and success.

Tame pastures also have weed problems that can usually be managed by herbicides, shredding, or cultivation. More details on weed control will be covered in Chapter 13.

Revegetation of rangeland is sometimes considered using native or introduced herbs and grass species. Replanting on arid rangeland is difficult due to competing vegetation and dry conditions. It is also equally difficult in the more humid regions because of competing vegetation. Therefore, some form of weed control may be necessary to allow the developing young seedlings of planted forage to emerge and survive.

Wildlife Management

Wildlife is usually an important component of rangelands and consists of small mammals, birds, and big game animals. These animals are mainly rodents, rabbits, deer, pronghorn, elk, big-horn sheep, mountain goats, wild turkey, ducks and various pheasants and game birds. Wildlife populations are regulated by the availability of food, water, and cover. Livestock use may benefit wildlife by opening up dense stands of vegetation in the more humid areas, and they are usually not a problem in wildlife habitat in arid areas unless overgrazed.

Heavy mule deer use in some areas of sagebrush country does not reduce cattle use of grasses, and wildlife use of vegetation is usually not affected. Vegetation diversity and cover favor wildlife. Carefully controlled livestock grazing can complement wildlife. Wildlife and livestock habitat use can be compatible if not overused since they utilize different plant species in many cases.

Predators such as coyotes influence livestock numbers and can kill sheep and goats. Other predators include black bears, golden eagles, bobcats, foxes, and mountain lions. They are also important predators on livestock. Grizzly bears and wolves are low in numbers so they cause little livestock loss. Rodents, rabbits, and prairie dogs may compete with livestock some by use of plants on rangelands. Wild horses and burros may also compete with livestock for forage on some western rangelands. Insects such as grasshoppers may damage plant cover and de-stroy forage that could be used by livestock or wildlife. Their populations fluctuate considerably from year to year as well as the Mormon cricket on some western rangelands.

Game birds, small and big game animals are a source of food and hunting. This subject is discussed further in Chapter 10 and the importance of hunting on many ranches and farms.

Animal Nutrition on Rangelands

Holechek et al. (2002) indicated nutrition for range animals is more difficult to define than confined animals because of the added energy required to move about. Also wind, heat, and extreme cold add environmental stresses.

Basically, carbohydrates, fats, proteins, minerals, and vitamins are obtained from plant cells they consume. Carbohydrates are composed of carbon, hydrogen, and oxygen and provide energy for range animals. Starches and sugars found in plant cells are readily broken down by the animals through the digestive process and provide energy. Cellulose and hemicellulose in plant cells can be broken down by ruminant animals (cattle, sheep, goats, bison, elk, giraffes, and camels) or animals with enlarged cecums (horses and rabbits) because they possess microorganisms capable of digesting these carbohydrates. Therefore, digestion of cellulose and hemicellulose is a slow process compared to starch and sugars. Lignin found in plant cells cannot be utilized by animals because it is resistant to breakdown by microorganisms. Lignin is a main component of woody plants.

Fats have more energy than carbohydrates and serve as stored energy for the animal. Typically range plants are low in fat.

Proteins contain nitrogen in addition to carbon, hydrogen, and oxygen. Proteins function as enzymes, hormones, antibodies against diseases and as agents for transport and storage of nutrients in the range animal. Proteins are not stored in the animal so a continuous source is required. Actively growing plants have a much higher protein content than dormant plants. During the dormant season, grazing animals may need supplemental crude protein or be moved to other locations that provide needed protein and proper nutrition.

Minerals needed for range animals include calcium, phosphorus, sodium, potassium, magnesium, chlorine, and sulfur.

Phosphorus is the most limiting of grazing animal productivity. Microminerals for grazing animals include iron, iodine, copper, cobalt, fluorine, zinc, molybdenum, selenium, and manganese in minute quantities (<0.01% of body weight). Also trace amounts of silicon, tin, vanadium, and nickel have been reported as essential. Selenium is the micronutrient if too low can cause white muscle disease in calves and lambs. If selenium levels are too high in forage, it can be toxic.

Vitamins are either fat-soluble (A, D, E, K) or water-soluble (C and B complex). Fat-solubles are stored in animals for use during periods of lack. Water-soluble vitamins, with the exception of B2 and B12, are generally not stored and a consistent supply is needed.

Forage intake of grazing animals varies with body weight as well as forage quality and availability. Most range ruminants require about 2% of their body weight per day but it varies with season. Horses consume about 60 to 70% more forage than ruminants when quality of forage is comparable.

Range plants vary in crude protein, minerals, cell soluble levels, fiber, and lignin content. Leaves are higher in nutrients than stems. Fruits and flowers from forbs and shrubs have higher cell solubles and crude protein usually than leaves. Buds of shrubs are particularly high in cell solubles and protein. Seeds from grasses are higher in protein and cell solubles than in leaves. Plants during dormant periods are lower in nutrition than when actively growing, and palatable forage species vary considerably in nutrients (proteins, starches, and sugars). Animal nutrition on rangelands is extremely important and complex and the reader is referred to the literature cited (Holechek et al. 2002; Hanselka et al. 2014).

Multiple Use of Rangelands

Some discussion was presented earlier in this chapter on products of grazing lands. In addition to cattle, sheep, goats, and horses, many of the federal rangelands support wild horses and donkeys. Care must be taken on those lands not to overgraze and cause soil erosion and degradation of forage and native vegetation.

On public lands some off-road vehicles, motorcycle or recreation vehicles may be used and care must be taken not to cause excessive damage to soil or vegetation. It is prohibited on much of the federal lands.

Rangelands in many areas provide watersheds (like forested areas discussed in Chapter 7) and water use for urban and local domestic use. Water harvesting methods (water cycle discussed in Chapter 3) from runoff is captured by various methods to improve water availability to the range plants and wildlife. These include making ditches and furrows, pitting, ripping, chiseling, water spreading, or harvesting to direct water where needed on the rangeland.

Timber production and mining also occur on rangelands where appropriate, and rangelands also contribute much oil and gas production.

Rangelands provide horseback riding, fishing, hunting, camping and hiking opportunities and scenic beauty. In addition, rangeland provides space for wildlife and endangered plant and animal species not provided elsewhere.

Rangelands add billions of dollars to our national economy and welfare in water storage and use; animals, plants and wildlife sanctuaries, wood products, minerals, human habitation, fossil fuels, a natural sink for CO_2 sequestration, and space for future needs and expansion.

Pasturelands and Hay

Pastures can be defined as about any land used by grazing domestic animals. It differs from rangeland in that periodic cultivation is used to maintain introduced non-native forage species and agronomic inputs such as irrigation and fertilization are sometimes used. Heath et al. (1975) defines pastures as fenced areas of domesticated forages, usually improved, on which animals are grazed.

Plants for Pasture and Hay

The principal forages on pastures are largely in two botanical families, the grasses, Gramineae, and the legumes, Leguminosae (see Chapter 4). The grasses are grouped into about 600 genera with nearly 5000 species. About 1500 species of 150 genera are found in the United States. The grass family includes about 75% of the cultivated forage crops and all cereal crops (Metcalfe 1975).

The legumes have nearly 500 genera with about 11,000 species of legumes worldwide, with about 4000 in the United States. The stages of growth of grasses and legume crops and their improvement are explained in Chapter 4.

The major rangeland and pasture grasses in the U.S. include the cool-season wheatgrasses and ryegrasses and warm-season grasses such as big bluestem, Indiangrass, switchgrass, little bluestem, sideoats grama, weeping lovegrass and bermudagrass. Other cool-season grasses include tall fescue, orchardgrass, timothy, smooth bromegrass, and reed canarygrass.

Cool-season legumes include alfalfa, red clover, white clover, and birdsfoot trefoil as well as numerous annual clovers and medics. Alfalfa is grown extensively worldwide and in the U.S. for pasture and hay. Red and white clover is used for hay, pasture, green manure, silage, and as a cover crop. Birdsfoot trefoil is native to Europe, north Africa, and Asia and is used for pasture, hay, silage, soil improvement, and road bank stabilization.

The cool-season grasses Italian and perennial ryegrass are used for pasture and hay in the Pacific Northwest and also for silage and turf grasses. Tall fescue, also introduced from Europe, can be used for hay, pasture, turf, and erosion control. Orchardgrass, introduced from Europe, is a long-lived perennial bunch type grass for pasture, hay, silage, cover crop, and ornamental. Smooth brome is used in soil erosion control, pasture, and hay. Reed canarygrass can be used for silage and hay. Seeds are used in bird feed.

Breeding switchgrass as a forage crop has occurred since the 1950s mainly to improve biomass for ethyl alcohol production, although it is good for pasture and hay use.

Morphology of Grasses

Grasses are either annuals or perennials. Almost all are herbaceous (nonwoody plants) (Metcalfe 1975). The grasses are monocotyledons. A cotyledon is the first leaf or leaves of the embryo in seed plants. Grasses have one cotyledon and are distinguished from legumes which have two cotyledons in the embryo.

Grasses first have vegetative growth followed by stem elongation, boot formation, heading and flowering, seed development, and seed and stem ripening. Vegetative regrowth may occur if tops are removed by grazing, fire, or frost during and late in the growing season. If heavily grazed by animals, the plant could remain in the vegetative stage during the entire growing season. The growing point or meristematic tissue is protected by a surrounding leaf sheath making the plant resistant to damage by fire or grazing animals. Some perennial grasses have creeping stems called stolons or rhizomes (horizontal underground stems) that support, store food reserves, and help in overwintering of the plant. The leaves are borne on the stem, alternately in two rows, one at each node. The stem is divided into nodes and internodes. The internode may be hollow, pithy, or solid. Each node is solid. Lateral buds can arise in the axis of the leaves and become vegetative branches of the stem or flower shoots. Brace roots occur in corn and some grasses and arise from the nodal meristem, a zone just above the node that helps keep the plant erect. Roots and seed production in forage grasses are similar to those reported for the grass crops in Chapter 4. Grasses have fibrous root systems, and extensive roots develop from the lower nodes of the young stem. Secondary roots sometimes form at nodes above the ground or at the nodes of creeping stems.

The flowering part of a grass is called the inflorescence and is made up of smaller units called spikelets. Spikelets are usually grown in groups or clusters of several types, including spikes, racemes, and panicles. Wheat and barley have spikes. Smooth bromegrass, Kentucky bluegrass, and red top are panicles.

Leaves of grasses are borne on the stem, alternately into rows, one at each node (Metcalfe 1975). The leaf consists of a sheath, blade and, when present, a ligule. The sheath surrounds the stem above the node. The blades are parallel veined and typically flat, have oracles or ear-like appendages projecting from the leaf edges at the junction of the sheath and blade. The ligule is the appendage that clasps the stem where the sheath and blade join.

Morphology of Legumes

Leaves of legumes are arranged alternately and usually have large stipules at their base and are either pinnately or palmately compound. Stems of legumes vary greatly in different species in length, size, and amount of branching and woodiness. The roots are usually taproots and nearly all are associated with N-fixing bacteria that replenish nitrate in the soil.

Flowers are usually arranged in racemes as the pea, in heads as in clover, or in a spikelike raceme as in alfalfa. They are usually crossed-pollinated. The fruit is a pod containing one to several seeds. The seed is usually without an endosperm at maturity but reserve food is stored in the cotyledons. Each seed is enclosed in the testa or seed coat.

Management of Pastures

Management of a pasture may be more intense than a rangeland because they usually require more management for weed control, fertilization, and irrigation. Weeds and woody plants may be a problem in pastures and sometimes need to be controlled by prescribed fire, shredding, mechanical means, or herbicides. Many times the weed problem is ignored and some grazing of these plants may be possible with livestock.

Pastures may also occur on fertile soils with adequate water and favorable climate for livestock use. Some pastureland, such as bermudagrass or alfalfa, is used to produce silage and hay crops.

As indicated in Chapter 4, the cool-season forages (legumes and grasses) and the warm-season forage grasses have been improved in yield, vigor, and nutritional features by breeding programs and cultural practices.

Some crops may be pastured periodically during the year or intermittently in separate years by cleaning up plant residues after harvest of field or hay crops, pasturing of young wheat plantings, temporarily pasturing hay fields or some other situation to consume available forage. Nutritional requirements are similar to those reported for livestock on rangeland and method of livestock and stocking rate is dedicated by choice and resources. Wildlife use should also be considered for the resources and situation.

Pasture lands may need to be renovated periodically by re-seeding or revegetation with a new or different plant cultivar, fertilized, irrigated and/or controlled for weeds. All these activities require a lot of planning, time, expense, and hard work. Like all farming and ranching enterprises, great effort, time, and resources are needed to be successful.

Summary

Rangeland is defined as all areas of the world that are uncultivated land such as deserts, forests, and all-natural grasslands. Rangelands provide grazing for browsing animals and many other activities. Rangelands are distinguished from pasturelands in that pasturelands require agronomic inputs to maintain introduced non-native forage species.

Grazing lands occur on about 26% of land area in the United States (2.3 billion acres), an area nearly equal to forested land.

Products of grazing lands are mainly wool and red meat (cattle, goats, sheep) and provides space for a tremendous variety of wildlife. Grazing lands provide huge water reserves, recreation, human habitation, wood products, minerals, scenery, renewable, and nonrenewable energy. It serves a huge sink for CO_2 sequestration and environmental maintenance.

The problems of grazing lands include uncertain rainfall and weather (drought), overuse and desertification, poisonous plants, weeds, and predators (coyotes, wolves, rodents, insects).

The types of rangeland in the United States includes grassland, desert shrubland, savanna woodland, forest, and tundra. In the United States about 15 different rangeland categories occur within those listed and each is described by soil, vegetation, rainfall, and location.

Management of rangelands is assessed by taking inventory of the vegetation, calculating the livestock stocking rate, grazing method, wildlife management, and nutritional needs of the livestock and wildlife for each area.

Pasturelands and hay are produced in a variety of locations, land, and agronomic inputs. The main grasses and legumes used are described by morphology and name. Management of pastures may be more intense than rangelands because plants may require more weed control, fertilization, and/or irrigation than plants on rangelands for maximum productivity. Grasses and legumes grown for pasture and hay crops have been improved in nutrition and vigor by breeding programs, cultural methods, and new technologies.

Chapter 8. Literature Cited

Lubowske RN, Vesterby M, Bucholtz S, Baez A, Roberts M (2006). *Major Uses of Land in the United States*, 2002. United States Department of Agriculture, Economic Research Service. Economic Information Bulletin No. (EIB-15) 54 p.

Holechek JL, Pieper RD, Herbel CH (2002). *Range Management: Principles and Practices.* 4th Edition. Custom Edition for Texas A&M University. Prentice-Hall, Inc. Upper Saddle River, NJ. 571 p.

Wooten HH, Bernes CP (1948). A Billion Acres of Grasslands. Pages 25-34 in Stefferud A. ed. Trees. The Yearbook of Agriculture 1948. United States Department of Agriculture. U.S. Government Printing Office, Washington D.C.

Lark TJ, Salmon JM, Gribbs HK (2015). Cropland Expansion Outpaces Agricultural and Biofuel Policies in the United States. Environmental Research Letters 10 (2015) 044003.

Holechek JL, Sawalhah MN (2014). Energy and Rangelands: A Perspective Rangelands 36:36-43.

Holechek JL, Sawalhah MN, Cibils AF (2015). Renewable Energy, Energy Conservation. and U.S. Rangelands. Rangelands. 37 (6) 217-225.

Hancock A (2006). Doing the Math: Calculating a Sustainable Stocking Rate. North Dakota Agricultural Experiment Station and Central Grasslands Research Extension Service. Home 2000 Annual Report. Retrieved from https://www.ag.ndsu.edu/.../Doing%20the%Math.htm. North Dakota State University. Accessed 6 June 2016.

Hanselka W, Lyons R, Moseley M (2010). *Grazing Land Stewardship: A Manual for Texas Landowners*. AgriLife Extension, Texas A&M System. B-6221. 165 p.

Heath ME, Metcalfe DS, Barnes RF (1975). *Forages: The Science of Grassland Agriculture*. 3rd Edition. The Iowa State University Press. Ames. Iowa. 755 p.

Metcalfe DS (1975). The Botany of Grasses and Legumes. Pages 80-97 in Heath ME, Metcalfe DS, Barnes RI, eds. (1975). Forages: The Science of Grassland Agriculture. 3rd Edition. The Iowa State University Press. Ames, Iowa.

LIVESTOCK AND POULTRY

Introduction

Domestication of animals (pigs, sheep, goats, and cattle) occurred about 6500 to 7000 B.C.E. (see Chapter 1). Planting and harvesting of cereal grains occurred sometime between 7700 to 8000 B.C.E. (Levack et al. 2007). The first signs of goat domestication occurred about 8900 B.C.E. in the Zagros Mountains in Southwest Asia. Domesticated cattle, goats, and sheep occurred around 6500 B.C.E. and pigs about 7000 B.C.E.

Thompson (2010) indicated different people in different places domesticated different plants depending on what they found around them and the growing conditions they faced. The Soviet geneticist Nicholai Vavilov (1887-1943) was a pioneer in documenting the eight centers in the world of crop origin and diversity. They are China, Southeast Asia and the Pacific Islands, Central Asia, Near East, Mediterranean, Ethiopia, Mesoamerica and South America, and the Andes.

Damron (2009) indicated the earliest domestication of sheep was somewhere around 8000 B.C. followed closely by goats, pigs, and cattle at around 6500 B.C. as suggested by Levack et al. (2007). Llamas, horses, donkeys, and chickens were domesticated about 5500, 3500, 4000 and 6000 B.C., respectively.

I have indicated domestication of animals occurred about the same time as crops since domesticated animals need abundant plant products to grow and prosper. Plants and seeds of all types provide food for grazing and confined, wild, and domesticated animals so the plant-animal system complemented each other and allowed civilization to progress. Today we "take for granted" the plant-animal interactions that provide our milk, eggs, clothing, shelter, meat, wool, and other products used daily.

As a young boy some 80 years ago, we raised beef cattle, hogs, chickens, and also had a milk cow in rural Idaho in the 1930s through the 1950s. Wheat, barley, and oat crops were raised to sell and to feed the livestock. All meat, eggs, and milk products were used at home but the cattle and hogs were raised mainly for income. We also raised a garden.

Many rural people still produce their own food, but in 2024 most people purchase food products from the grocery store. I am sad to say many people in the United States do not have a clue where their food comes from or the tremendous effort it takes to produce it; hence this book to help explain the need for continued research, teaching and education in this area. The world population is expected to be 9 billion by 2050 (Buchanan 2016) (see Chapter 1). In order to feed this many people for a quality life, agricultural products need to be continually monitored and updated to satisfy demand.

Improved crops and animals are continually needed to improve vigor against diseases, insects and other pests, and to improve nutrition and quantity. This is because pests find new ways to attack crops and animals through their own biological adaptations. If agricultural research goes lacking, agriculture output will suffer as well as our quality and quantity of life. Understanding soils and soil availability, water use and supply, nutritional requirements of plants and animals, and management, markets and government policies are necessary and significantly affect agriculture output. The United States does a good job of research and education, but the shrinking number of farmers and ranchers and those trained to do research and promote agriculture is of concern.

Feeding the World: Agricultural Research in the Twenty-First Century by Dr. Gale Buchanan is a book that outlines the history and future needs of agricultural research. It is a must-read for anyone interested in agriculture and our perilous food supply (Buchanan 2016).

Economic Benefits of Livestock and Poultry

Dillivan and Davis (2014) reported that the U.S. animal agriculture impact to the domestic economy in 2012 provided nearly 2 million jobs, $346 billion in economic output, $60 billion to household income, generated $15 billion in paid income tax and $6 billion in paid property tax.

They also reported increased production from 2002 to 2012 at about 53%, 32%, 22%, and 17% for milk, pork, turkey and eggs, respectively, in the United States. The authors indicated the impact of the animal agriculture during 2002 to 2012 produced 25 million additional dollars from animal products.

This data highlights the importance of the animal industry and its relation to crop production (feed for animals) and the economic benefits to humanity.

Animals of Concern

Horses

History

Urban and agriculture growth during the nineteenth and early twentieth centuries would have been impossible without horses and mules (McMillen 1981). The early development of horses is obscure, but the early Spanish and English adventurers brought horses to America. Whether escaped or stolen, they became mounts for the Indians. Millions of horses were used to plow and pull wagons and farm machinery. They were used to transport people, and farmers depended on them to cultivate fields and plant crops.

My grandfather and his family farmed with horses in the late 1800s and early 1900s in northern Idaho and raised wheat, oats, and barley (small grains). They kept about 50 draft horses and farmed about 1000 acres. By the time I was old enough to help, horses had largely been replaced by farm tractors (1940s). The Amish communities in the United States still use horsepower, as well as other people that just like draft horses. Mules were used as draft animals especially in the South (Buchanan 2016).

Breeds of Horses

The ranch horse of Texas and the western cattle country was selected for endurance and bursts of speed. The ranch horse had to carry the cowboy day after day on long cattle drives. Many were quarter horses.

Other breeds are the Albino, Appaloosa, Arabian, Morgan, Pinto, Sadler, Tennessee Walking horse, and the golden-coated Palomino with its silvery mane and tail (McMillen 1981). Horses were an essential part of urban and agrarian development in the

United States, but by 1950 there were more tractors on farms than horses.

Modern Times

In recent years the turnaround in horse numbers continues but more recreation horses are enjoying great numbers for riding enthusiasts, competitions, and for pets.

There was an estimated 20 million horses in March 2015 in the United States (Anonymous 2016a). A USDA census showed 4.5 million in 1959 but has dramatically increased since that time to about 20 million.

After domestication about 3500 years ago, horses were first used for food then for war and sports and also for draft purposes (Taylor and Field 1998). They were used to transport people and move heavy loads. In addition, horses became important in mail delivery, farming, forest harvesting, mining, and handling of livestock. Today horses are used in racing, handling of livestock and for companionship, recreation, and exercise.

When horses have served their usefulness and are no longer productive, they can be slaughtered and consumed by humans and pets.

Beef Cattle

History

Christopher Columbus was first to bring cattle to the Western Hemisphere and West Indies in 1493. Cortez brought cattle to Mexico in 1519 and Spanish missionaries spread cattle across the American West at the beginning of the 17th century (McMillen 1981; Damron 2009). The Spanish Longhorn was one of the livestock species at the Missions. In 1609 English settlers brought cattle to New England. The cattle were used for work and milk.

As settlements proceeded, the grasslands became increasingly used as beef-producing areas.

By the middle of the 19th century, beef cattle in the West were being improved with imported Shorthorns (Damron 2009). Shorthorns failed to meet the demand of range life and were crossed with Herefords and dominated the Western beef industry until devastated by a bad winter in 1886 when many cattle died. Later Angus crossed with Herefords were introduced (called Black Baldie) and are still part of the beef cattle industry.

After World War II, the next major shift in the U.S. cattle industry occurred (Damron 2009). Huge grain surpluses far removed from the cattle markets made shipping cost prohibitive, so enterprising cattle and grain producers began feeding cattle on surplus grain. This produced improved flavor and tenderness of the grain-fed beef. This became a finishing phase just before slaughter.

Range cattle fed in feedlots flattened too fast and didn't grow in size the industry wanted, so starting in 1965 other beef breeds were imported into the U.S. Total cattle numbers peaked in 1975 with 132 million head, including beef and dairy animals. The current number is about 95-105 million (Damron 2009). Total cattle numbers in 2015 were about 98 million and beef cows numbered about 30 million in 2016 (Anonymous 2016b).

Per capita consumption of beef in the United States was 53.84 pounds in 2014. It was 123.51 pounds in Hong Kong (highest in the world) and less than 1 pound in the Congo. Argentina, Uruguay, and Brazil consume more beef per capita than citizens of the U.S. (Cook 2014).

Major U.S. beef cattle breeds are Angus, Beefmaster, Brangus, Charolais, Gelbrich, Hereford, Limousin, Maine-Anjou, Red Angus, Shorthorn, and Simmental (Damron 2009).

The Beef Cattle Industry

The major segments of the beef cattle industry are seed stock producers, commercial cow-calf producers, yearling or stocker operations, and feedlot finishing operations (Damron 2009).

The primary goal of the seed stock industry segment is to produce breeding stock. Stock may be produced by purebred or controlled crossbreeding programs that produce the parent generation of cattle for commercial calf producers. The product of greater demand is bulls that are purebreds. Commercial cow-calf operations rely on these breeders for quality sires to mate with commercial cow herds. Purebreed seed stock producers can sell seed stock to other purebred producers if high quality and also market semen and embryos.

Commercial cow-calf operations represent the first phase of producing animals primarily for meat. The product is 6- to 10-month old, 300- to 700-pound calves sold at weaning to either a feedlot or a stocker calf operation. Cow-calf producers generally produce crossbreed calves bred for slaughter. Most U.S. cows calf in spring. Calves are usually marketed and the operator plans for next year's calf crop. Heavier calves may be placed in feedlots before marketing them.

Yearling or stocker operations grow calves to heavier weights on low-priced forage before the calves enter the feedlot. They typically purchase calves from cow-calf producers and grow them during a specific season and then ship them to feedlots. The heavier calves can also go directly to the feedlot, depending on grain prices or availability of pasture or other feed (hay or silage).

The feedlot phase is the finishing phase of beef production. Both steers and heifers are fed in feedlots. Feedlots vary in size and location. Some are small operations to huge that finish cattle for slaughter. Most calves are fed grain in feedlots before slaughter and processed in meat packing plants.

Texas, Kansas, and Nebraska are top beef feedlot and marketing states. Genetic and breeding programs are important to improve quality and quantity of beef and managing the nutrition, as explained in Chapter 8, is critical. High-quality protein feeds are necessary for feedlot cattle.

The goal of breeding herds of beef cattle is to produce and raise one calf per year from each mature female in the herd. The calf must be born at the right time of year for the operation's resources to generate profit. Good management skills are required to care for the cows and bulls to provide good nutrition and prevent and recognize any disease, pest, reproductive, or physical problems and treat them.

Beef cattle products include leather, many retail cuts of meat and blends with many other products such as pizza.

Dairy Cattle

Milk Production

Milk is regarded as the perfect food and has helped many youngsters grow into adults with good health and vigor (McMillen 1981). The tremendous flow daily of milk from cows to consumers (humans) is considered a marvel and part of our blessed heritage.

The world's population obtains most of its milk and milk products from cows, water buffalo, goats, and sheep with minor amounts from horses, donkeys, reindeer, yaks, camels, and sows but in the U.S. most comes from cows (McMillen 1981; Taylor and Field 1998).

Milk is a colloidal suspension of solids in liquid. Fluid whole milk is approximately 88% water, 8.6% solids-not-fat (SNF), and 3-4% milk fat. The SNF is total solids minus the milk fat. It contains protein, lactose, and minerals (Taylor and Field 1998). Milk is composed of as many as 500 different fatty acids and fatty acid derivatives, lactose (carbohydrates), about 3.3% protein (all amino acids required by humans), all essential vitamins, a rich source of calcium, and a good source of phosphorus and zinc.

Milk composition is markedly different for different mammalian species. For example, milk from reindeer and aquatic mammals is very high in total solids, whereas mare's milk is low. Milk of donkeys is low in fat (1.4%), and high in fur seal milk (53.3%). Milk from cows and goats is similar to human milk except for protein which is much higher in cows and goats (Taylor and Field 1998).

The Dairy Industry

The dairy industry is a modern marvel making fresh safe milk available to many people, even in crowded cities, because of the huge and high efficiency of the modern dairy, improved and ever-present transportation system, refrigeration, and pasteurization process (McMillen 1981).

Milk in its raw state is highly perishable and can easily be dangerous. Before modernization, vessels which conveyed or received it could contain or contaminate milk with airborne contagions of fevers, diphtheria, septic throat and other ills. The importance of prompt chilling and use of clean containers was necessary.

For centuries the only dairy item that was a commerce product was cheese, and it was used in colonial times. George Washington's army records disclose his acknowledgement and thanks for the gifts of cheese (McMillen 1981).

In the early history of the United States, cattle were cattle and no distinction between beef and dairy breeds were made. Milk was obtained from any cow that could be persuaded or compelled to give milk. To bring one large animal from England to the United States in colonial times cost five times as much as to bring one human being, but cattle were essential to the new settlement (McMillen 1981).

The dairy farmers always have had to work hard and long for income. The cows are milked twice a day, seven days a week; and best yield is obtained when milked at regular and equally spaced intervals. Modern dairy farms use certain breeds of dairy cattle. Most are not pedigreed animals. In the United States the Holstein breed is most abundant by far followed by the Jersey. Minor breeds are the Brown Swiss, Guernsey, Aryshire, Red and White and Milking Shorthorn. Great strides have been made over the past 50 years in improving milk production through improved management and breeding. Even so, the dairy farmer is confined to his work. It is difficult, unpredictable, and making a profit is sometimes wanting (Taylor and Field 1998).

Products of milk include fluid milk and cream, cheese, butter, ice cream, evaporated and condensed milk, fermented products (buttermilk, yogurt, sour cream) and others.

Managing Dairy Cattle

A large investment in cows, machinery, barns, and milking facilities is necessary. Some farmers grow their own feed; so additional machinery and labor to plant, harvest, and process the crops are needed. In addition, veterinary services to monitor and control diseases are necessary.

The average milk production (lactation period) per cow in the United States of 305 days is about 16,400 lbs. Consequently, a cow weighing 1400 lbs that produces 40 lbs of milk per day needs 1.25 times as much energy for lactation as it needs for maintenance. If producing more than 40 lbs of milk per day, she will need more energy for maintenance. For the first 2 to 4 months after calving, it is difficult to provide adequate nutrition because milk yield is high and nutrition intake is limited. The cow will use body fat and protein reserves to make up the difference.

The cow should be monitored and different feeds can be used to provide the needed energy, protein, vitamins, and minerals. Alfalfa hay, corn-oat grain mix, soybean oil meal, dical, salt, vitamins, and trace mineral mix would be a typical ration during the critical 2-4 month period.

Dairy producers should plan for a 50-60 day dry period (non-lactating female) to allow the cow to adequately improve body condition and mammary tissue before further milk production. Dry cows should be separated from lactating cows so they can be fed and cared for (Taylor and Field 1998).

Other considerations in the dairy operation are nutrition and replacement of heifers, nutrition of bulls, calving operations, adequate milking and housing facilities, milking operations, controlling diseases, and making a profit.

Hogs (Swine)

History

Columbus on his second voyage in 1493 to America brought eight pigs that populated the West Indies. Cortez carried pigs to Mexico from Cuba and Spain in the 1520s, and de Soto brought pigs to Florida in 1539 that apparently spread across the southern states. The English landed pigs in Jamestown in 1609, which rapidly spread by 1627. Settlers in the New World had pork and venison. Pork became commonplace for the colonists in addition to fish, tobacco, and lumber. Hog meat became a major item of trade. Pioneers moving west carried hogs and other livestock along the Ohio River and its tributaries (McMillen 1981).

By the 19th century hogs were driven eastward by the thousands from the valley farms to the seaboard cities in the United

States. Farmers had pigs to sell, and growing towns and cities wanted meat. Exports to Europe were also made.

By the mid-1850s Cincinnati packers reached their maximum with nearly a half million hogs. Thereafter, the center of production shifted north and west. Major producing states were Tennessee, followed by Kentucky and Ohio. Illinois became number one in packing, and Chicago became hog butcher of the world (McMillen 1981).

Improvements in hogs were made by farmers by crossing different breeds. Slave traders brought red pigs from the Guinea coast of Africa and red hogs were brought from other sources. This resulted in the Jersey red, and later as the Duroc Jersey, and is known now as simply the Duroc. Early breeding stock from diverse origins resulted in the Chester White, Duroc, Poland China, Berkshires, and Hampshires. Raising hogs in the Corn Belt allowed many farmers to get debt-free since hogs could bring more money than the corn crop. Hogs did very well on corn as feed.

Hog cholera unhappily found its way into the corn states as early as the 1860s and killed thousands of hogs. The cause of the disease was unknown but by 1908 Marion Dorset of the USDA Bureau of Animal Husbandry developed the serum to prevent cholera and saved the American farmers countless pain and money.

Breeders, aware of the changing market in the early 1900s, began to change the conformation of the pig from an obese prize hog to a meat type rather than the lard type. This produced less lard, better bacon, smaller and better hams.

Swine are intelligent and can take care of themselves. They quickly adapt to their environment and in nature are clean. They roll in mud to cool. They can be confined when properly cared for and can survive when given freedom. The hog will probably never become an endangered species because of their adapt-ability (McMillen 1981).

Swine Breeds

Major United States breeds of swine include Berkshire, Chester, Duroc, Hampshire, Landrace, Poland, Spotted, and Yorkshire (Taylor and Field 1998). The three most popular breeds are the Yorkshire, Duroc, and Hampshire. My father raised Duroc in the 1940s and '50s in rural northern Idaho when I was a small boy. He raised wheat, barley, and oats for feed and indicated he made more profit on the hogs than the wheat in the marketplace at that time.

The Swine Industry

The swine industry is a large sector of U.S. agriculture. Hogs are the fourth most important money generator in food animal agriculture. The gross annual income is about $14.2 billion and the United States produces about 9% of the world's pork. Swine are monogastric and use only limited amounts of forage, but are the most efficient converters of grain to red meat of all livestock species (Damron 2009).

Swine operations are classified as 1) feeder pig production, 2) finishing purchased feeder pigs, 3) farrow-to-finish and 4) pure-bred or seed stock (Taylor and Field 1998; Damron 2009). Feeder pigs are produced and later grown to market weight. They are pigs (farrow to wean) that are generally weaned early in mod-ern operations and placed in a nursery where they can receive specialized care until ready for finishing. In the farrow-to-finish operation, a breeding herd is maintained. Pigs are produced and finished for market on the same farm. The purebred and seed stock operations are similar to farrow-to-finish except the sellable product is primarily breeding boars, gilts, or show pigs which may be purebred or controlled crossbreeds.

An integrated corporate production can generally be described as farrow-to-finish but often have their own seed stock production as well. Operations of brood sow facilities, nursery facilities, finishing sites, and boar locations are usually separated far enough apart for biosecurity reasons.

Management of the sow during farrowing and lactation and the baby pigs from birth to weaning are the most critical periods in a total swine management program.

Sow rations and piglet rations must have the proper combinations of amino acids, vitamins, and minerals, in addition to protein and energy, to assure adequate and cost-effective gains. Herd health and programs must be well planned to prevent reproductive diseases, assure adequate lactation, and to prevent scours (diarrhea) and other diseases of baby pigs.

Excellent management is reflected in lower costs and higher profits. Returns are reflected by prices for farrow-to-finish (castrated younger males) and to gilts (young female swine) that can fluctuate from year to year. Top managers have low cost production (especially low feed costs) when possible.

Sheep and Goats

History

Sheep and goats were probably domesticated almost 8000 B.C. and were probably the first food-producing animals to be domesticated. Sheep and goats were associated with humans as our globe was populated (Damron 2009). Christopher Columbus also brought sheep and goats to the Western Hemisphere in addition to cattle to the West Indies on his second voyage in 1493. Cortez brought sheep, goats, and cattle to Mexico in 1519.

Sheep were brought to the East Coast in 1609 by the English who settled in New England.

McMillen (1981) indicated sheep were first brought across the Mexican border into the United States in 1540 by the conquistador, Francisco Vasquez de Coronado. Some 5000 sheep were brought, but most were eaten or lost. A few may have survived in Indian possession. However, Coronado established a great ranch south of Mexico City where he had churros (open-fleeced, coarse-wooled, and bare-bellied sheep). The churros were a good source of mutton. Coronado came to the New World for gold. The Catholic priests that came with Coronado established missions in the Southwest United States and took sheep to please their converts. Eventually sheep spread to Texas, New Mexico, Arizona, and California and adjacent areas by the late 1600s.

Sheep ranges were not always peaceful because of Spanish conquests against Native Americans, but some sheep always survived. Demand for mutton and wool by miners and Indians of the Southwest gave sheep a new life.

Spain for centuries maintained a monopoly on fine wool from Merino sheep. Revenues from the wool probably paid for the Columbus expedition.

England had no Merinos, but she had her own sheep, her own woolen mills, and the ability to import wool for those mills. The intrusion of the American Colonies into this business was not favored by Britain's leaders. Even so, Massachusetts claimed 100,000 sheep in 1664. By 1808 when Napoleon controlled Spain, he made a deal with the American consul at Lisbon. William Jarvis purchased 3500 Merinos and sent them to New England where they were sold and scattered among numerous owners. Even with all the trade problems, sheep growing and wool manufacturing, by 1860 the United States had more than 22 million head of sheep in the Northeast. They eventually made their way westward (McMillen 1981).

The Sheep and Goat Industry

The total count of sheep and lambs in 1867 just after the Civil War in the U.S. was 46.3 million head. From 1867 until now, sheep numbers have passed through many cycled phases. Sheep numbers peaked in 1942 at 56.2 million (Damron 2009). Sheep numbers were 6.9 million in 2001, but were below 6 million by 2015 (Anonymous 2016c). The reasons for decline in sheep numbers were because of less demand for wool and lamb meat and increased difficulty in keeping reliable herders, and competition for public-owned rangeland and increased grazing fees and predator problems. Decreased government support and profits and land use for other enterprises (crops for biofuels) also contributed to their decline (Damron 2009).

Sheep and goats are closely related in the United States, both originating from Europe and the cooler regions of Asia. Sheep are distinguished from goats by the absence of a beard, less odor (males only), and glands in all feet. Goat horns spiral to the left while sheep horns spiral to the right (Taylor and Field 1998).

Millions of goats are concentrated primarily in India, China, Pakistan, Nigeria, and Bangladesh. Goats are important for milk and meat. The purpose of sheep and goats in the United States is to take advantage of forage and roughage to produce milk, meat, and fiber. However, the value of wool has declined to the point of being unprofitable. Synthetics and availability of wool on the world market have reduced the importance of its production in the United States (Damron 2009).

The Angora goat production is shrinking because of the lack of government payments for mohair. Meat production, however, has potential for expansion primarily in the South.

The dairy goat sector has had a few large-scale commercial operations and numerous small-scale operations who milk goats for small returns or as a hobby. Typically dairy farms are done with dairy cattle in the U.S.

The dairy sheep does well in many countries, especially those surrounding the Mediterranean. Dairy sheep has a rich history and produces some of the world's best and most expensive cheese for the gourmet market. Dairy sheep are in the development stage in the United States. Goat's milk also makes special gourmet cheeses.

Sheep and goats are spread across the United States. The western states have the majority of all breeding sheep. They are sometimes mixed with cattle on rangeland. Many sheep graze on Bureau of Land Management and Forest Service land belonging to the U.S. Government. Texas has three-fourths of the Angora goats in the United States on the Edwards Plateau (Damron 2009).

Sheep and Goat Breeds

Major U.S. sheep breeds include the Suffolk, Dorset, Rambouillet, and Hampshire with lesser numbers of Columbia, Southdown, Corriedale, Polypay, Montadale, Shropshire, and Cheviot (Taylor and Field 1998). Major goat breeds are grouped as milk, meat, dual-purpose, or fiber breeds.

Managing Sheep and Goats

Sheep are capable of producing more than one offspring per female per year. Most operations maximize output of lambs by increasing the number of multiple births and eliminating ewes that do not lamb. Since gestation in sheep is only 148 days, more than one lamb crop per year is possible with good care and nutrition. Disease prevention program is a must since sheep are sensitive to internal parasites.

In the goat, scent glands located around the base of the horn help stimulate estrus and improve conception. The odor can be unpleasant to humans and affect the flavor of the milk. For this reason, bucks are generally kept separate from the milking does in use for human consumption. Like sheep, many breeds of goats respond to length of daylight and are thus seasonal breeders. Sheep and goats reach sexual maturity at about 6 months of age. Length of gestation is about 150 days. The U.S. normal breeding time is August through February. Some breed twice a year, but this is more prevalent the further south the animal is found (Damon 2009).

Feed is the single most expensive part of sheep production. Cost must be contained so most feed is forages and roughage with limited amounts of grain. Goats can be produced without extensive feeding strategies, but dairy goats require higher-quality feed. Good quality forage supplemented with commercially-available dairy goat feed is preferred. Goat's milk is very similar to milk of dairy cows (Damron 2009).

Lamb consumption is about 1 lb per capita per year and has declined to a specialty industry. Consumption of goat meat is unknown in the U.S. but the trend is upward because of the expanding immigrant population. Domestic consumption of wool and mohair is essentially nil because of less expensive synthetics and cotton. There are several associations like cattle, horses, and swine that represent the sheep and goat industry (Damron 2009).

Poultry

History

Damron (2009) stated that poultry has been with us for many thousands of years. DNA testing indicated modern chickens are derived from the Red Junglefowl (*Gallus gallus*) native to Thailand and are still found in the wild today. Cockfighting has been a favorite sport of many civilizations. Domestic chickens were known in India over 3400 years ago. Both Egypt and China had domestic poultry by 1400 B.C. Columbus brought chickens on his second voyage in 1493 to the Western Hemisphere, and in 1607 chickens were brought by the Jamestown Colony settlers to North America. The flock provided meat and eggs and income from extra produce provided cash for household staples such as sugar, flour, coffee, and other commodities (McMillen 1981; Taylor and Field 1998; Damron 2009).

The commercial hatchery industry got underway after the incubator became available in 1844, and by the 1870s small home-sized incubators became available. With a kerosene heater and the temperature of 102.5°F, chicks can be hatched from the eggs by the end of three weeks. Commercial hatcheries got another boost in 1918 when the U.S. Post Office allowed chicks to be shipped by mail. This allowed the hatcheries to increase in size and improve their brood stock. They could ship to any location in the U.S. I remember as a boy in rural Idaho my parents ordered and received baby chicks from commercial hatcheries through the local U.S. Post Office in the 1930s and 1940s. The chicks arrived in good shape and were placed in good housing with supplemental heating to begin their journey. A mix of male and female chickens were received.

All animal agriculture, including poultry, received another boost in 1862 with the passage of the Morrill Agriculture Land

Grant Act. This allowed establishment of the extensive land grant system of colleges of agriculture in all states. The Hatch Act was passed in 1887, which established agricultural experiment stations and the ability to conduct research to study and improve crops and animals (Buchanan 2016). The final piece of legislation was the Smith-Lever Agricultural Extension Act which established the Cooperative Extension Service designed to distribute information to ranchers and farmers. The poultry and other agricultural industries have benefited tremendously from the research, education, and information available from the land grant university system.

The Poultry Industry

According to Taylor and Field (1998), chickens are classified according to class, breed, and variety. A class has been developed in the same broad geographical area. The four classes of chickens are American, Asiatic, English, and Mediterranean. The American breeds are White Plymouth Rock, Wyandotte, Rhode Island Red, and New Hampshire. A breed is a subdivision of a class composed of birds of similar size and shape. A variety is a subdivision of a breed with birds of the same feather color and comb type.

Factors such as egg numbers, eggshell quality, egg size, efficiency of production, fertility, and hatchability are most important to commercial egg producers. Broiler producers consider white plumage and picking quality, egg production, fertility, hatchability, growth rate, carcass quality, feed efficiency, and livability important. Also breeds, strains, and lines that cross well with each other are important (Taylor and Fields 1998).

The modern turkey is a descendant of the wild turkeys native to North and Central America. There are eight varieties of turkeys: Bronze, Narragansett, White Holland, Black, Slate, Bourbon

Red, Beltsville Small White, and Royal Palm. All were developed in the United States except the White Holland. Only two types are commercially important and include the Small White and the large Broad White.

The Mallard is the most popular duck in the U.S. The White Pekin duck native to China (brought in 1873) is the most popular breed. It is most valuable for its meat and produces excellent carcasses at 7-8 weeks of age.

Several breeds of geese are popular. The Embden, Toulouse, White Chinese, and Pilgrim are all meat producers. The domestic goose was bred in Ancient Egypt, China, and India and brought to the U.S. in colonial times probably via Europe. The Embden and White Chinese breeds are best laying, produce good carcasses, grow rapidly with good livability (low mortality) and have a heavy coat of white feathers or nearly white (Taylor and Field 1998; Damron 2009).

Management of Poultry

In general, the management of poultry is very much alike for all species. The breeder flock needs proper nesting facilities and ample clean nesting material. Eggs should be collected at least once a day and more frequent in commercial hatcheries. Hatching eggs cannot be washed because it removes protective seals, and cracked eggs should be discarded. Commercial operations fumigate eggs prior to setting to reduce bacteria and increase hatchability (Damron 2009).

Eggs should be placed soon in incubators or held at 60°F with 75% humidity before incubation. Cool temperatures delay embryonic growth until incubation begins and high humidity prevents moisture loss. Storage should be less than 10 days before incubation of eggs.

To produce hatching eggs, several mating systems are followed. Mass mating is where several males run with a flock of females. Pen mating is where one male is mated with a small flock of females or a system where one female is mated with one male then removed and another female is put with the male. This way more females can be mated to a superior male. A fourth way is artificial insemination.

Egg production is also accomplished in large units commercially and requires large capital investment. Environmentally controlled housing and computer technology are common. Small units also exist for the home grower for personal use, but in commercial production most eggs are never handled by humans. Geographically the poultry egg industry is located close to the human population. Eggs require less processing than broilers so markets easily move where eggs are sold (Damron 2009).

Meat chickens are marketed as broilers, roasters, and game hens in the broiler industry. Production is usually concentrated on large operations with large investment and facilities. The broiler industry is located primarily in the southern and southeastern states because of favorable climate, low-cost labor, and near large human populations.

Turkeys are typically reared on productive units that grow 50,000 to 75,000 birds per year. Turkey production is the third largest segment of the poultry industry following broilers and laying hens (chickens). Turkey production is year-round. Turkey hens are marketed between 14 and 16 weeks of age at 14 to 18 lbs. Toms are marketed between 17 and 20 weeks of age at 26 to 36 lbs.

Modern poultry breeding is a very controlled operation. All commercial poultry today are crosses of breeds, strains, or inbred lines. Chickens are either meat or egg types. Chickens exist in many colors, sizes, and shapes and in many combinations of these types. As indicated above, they are designated by class, breed, variety, and strain. The American Poultry Association maintains a standard to recognize breeds and varieties of chickens, bantams, ducks, geese, and turkeys (Damron 2009).

The poultry industry in the United States produces inexpensive sources of protein for humans. Poultry are efficient converters of grain and by-products into meat and eggs. Modern diets vary with growth stage and bird type but are similar in basic components. Commercial feeds are complete mixed diets that are often in mash or pellet form.

Disease control is essential because many diseases affect poultry. Biosecurity is protected and all equipment, coops, crates, and containers should be cleaned before and after use. Only essential workers should visit the flocks.

Eggs contain all essential amino acids needed by humans, many needed minerals, and all required vitamins except vitamin C. A large egg contains only 75 calories. Poultry meat products are high in protein and a good source of vitamins and minerals.

Value of the Poultry Industry

The poultry industry in the United States has been growing at a steady rate. The United States produces about 9% of the world's chicken eggs and 22% of the chicken meat and is the largest poultry meat producing country in the world. The U.S. is second to China in chicken egg production and is the largest turkey meat producer in the world.

The most recent census of Agriculture reported 233,770 poultry farms in the United States in 2012. In 2014 the U.S. poultry industry produced 8.54 billion broilers, 99.8 billion eggs, and 238 million turkeys. The combined value in 2014 was 48.3 billion dollars.

Other Animals

Other animals of importance include bees, fur bearers (mink, rabbits, fox), fish, dogs and cats and rabbits. These are minor numbers and the reader is referred to McMillen 1981; Taylor and Field 1998; Damron 2009 for further reading.

Summary

Domestication of animals (pigs, sheep, goats, and cattle) and plants occurred roughly about the same time, 6500 to 8000 B.C. As civilization progressed, plants and animals complemented each other since humans and animals depended on plants and plant products to survive. Humans depended upon both animals and plants to survive and humans have steadily improved and protected plants and animals for their own well-being.

Many people still produce their own food, or partially do, but most people in the United States purchase food from the grocery store and may not know where their food comes from or the tremendous investment and effort needed to produce it.

The world population is projected to be 9 billion by 2050 and in order to feed, clothe, provide shelter, and maintain a quality of life, responsible research, teaching, and extension is needed in agriculture.

Animal agriculture adds 2 million jobs and $346 billion to our economy and provides abundant meat, milk, and egg products to our table.

Horses have made colonizing and developing the United States possible as draft and work animals. Today they have largely been replaced by tractors and recreation vehicles. However, today there are an estimated 20 million horses in the United States largely for recreation, racing, and pets but many are still used in handling livestock. There are several breeds.

Beef cattle were first brought to the United States by the early Spanish and English adventurers. By the middle of the 19th century, beef cattle in the west were being improved with imported Shorthorns. Shorthorns failed to meet the demands of range life and were crossed with Herefords until devastated by cold weather in the late 1800s. Later, Angus crossed with Herefords were called Black Baldies and are still part of the beef cattle industry. After World War II, huge grain surpluses made feedlots possible and other beef breeds were imported and bred for the market. Major beef breeds are Angus, Beefmaster, Brangus, Charolais, Gelbvich, Hereford, Limousin, Maine-Anjou, Red Angus, Shorthorn and Simmental. There are about 95 to 105 million head of beef cattle and dairy cattle in the U.S. and Americans consume about 54 pounds of beef per capita.

Dairy cattle provide the perfect food in milk. Milk is also obtained from water buffalo, goats, sheep, horses, donkeys ,and reindeer especially in other parts of the world. The dairy industry is a modern marvel making fresh, safe milk available to most people. In the U.S. the Holstein breed is most abundant followed by the Jersey. Milk has many fatty acids, lactose (carbohydrate), protein, all essential vitamins, calcium, and phosphorus. No distinction between beef and dairy cattle was recognized in our early history, but eventually the modern dairy cow emerged and great strides have been made in improving milk production the past 50 years.

The many products from dairy include fluid milk and cream, cheese, butter, ice cream, evaporated and condensed milk, buttermilk, yogurt, sour cream, and others.

Columbus, on his second voyage in 1493, brought pigs, goats, and cattle to the West Indies as did Cortez to Mexico and de Soto to Florida in the 1500s. The English brought pigs to Jamestown in 1609. Pigs flourished in early America (U.S.A.)

and are still so today with many wild hogs roaming the country. New breeds have moved the hog from an obese prize pig to a meat type rather than the lard type producing more bacon and better hams. The most popular breeds are the Yorkshire, Duroc, and Hampshire.

Sheep and goats were probably domesticated about 8000 B.C. and the first food-producing animals to be domesticated. Sheep and goats were associated with humans as our globe was populated. Spain, for centuries, maintained a monopoly on the Merino sheep which produced fine wool; and it wasn't until about 1800 they were purchased and brought to the U.S. The total count of sheep in 1867 just after the Civil War in the U.S. was 46.3 million and peaked to 56.2 million in 1942. However, today there are less than 6 million head. This is due to less demand for wool and lamb meat, few reliable herders, decreased government support, and other factors. Goats are also less popular for the same reasons but a few large-scale commercial operations do exist. Most of the Angora goats in the U.S. are in Texas on the Edwards Plateau. The main sheep breeds are the Suffolk, Dorset, Rambouillet, and Hampshire. The main goat breeds are grouped as milk, meat, dual-purpose, and fiber breeds.

Poultry includes chickens, turkey, geese, and ducks. Chickens are produced for meat and eggs; and turkeys, geese, and ducks are mainly meat producers. Management is similar for chickens, turkeys, geese, and ducks. The egg is an ideal food and has all essential amino acids. many minerals, and all required vitamins except Vitamin C. Poultry meat is high in protein and a good source of vitamins and minerals.

Chapter 9. Literature Cited

Levack B, Muir E, Veldman M, Maas M (2007). *The West: Encounters and Transformations.* 2nd Edition. Pearson Longman, New York, NY 957 p.

Thompson P (2010). *Seeds, Sex and Civilization: How the Hidden Life of Plants Shaped Our World.* London, Thames and Hudson, 272 p.

Damron WS (2009). *Introduction to Animal Science: Global, Biological, Social and Industry Perspectives.* 2nd Custom Edition for Texas A&M University, Pearson Education, Inc. Prentice Hall, Upper Saddle River, New Jersey. 841 p.

Buchanan GA (2016). *Feeding the World: Agricultural Research in the Twenty-First Century.* Texas A&M University Press. College Station, TX. 322 p.

Dillivan K, Davis J (2014). Economic Benefits of the Livestock Industry. iGrow. A Service of South Dakota. Retrieved from igrow.org/livestock/profit-tips/economic-benefits-of-the-livestock-industry/. Accessed 4 July 2016.

McMillen W (1981). *Feeding Multitudes: A History of How Farmers Made America Rich.* The Interstate Printers and Publishers, Inc., Danville, IL. 489 p.

Taylor RE, Field TG (1998). *Scientific Farm Animal Production: An Introduction to Animal Science.* 6th Edition. Prentice Hall, Upper Saddle River, NJ. 702 p.

Anonymous (2016a). Horses in the United States. Wikipedia. The free encyclopedia. Retrieved from https://en.wikipedia.org/wiki/Horses_in_ the _United _States. Accessed 5 July 2016.

Anonymous (2016b). United States Department of Agriculture. National Agricultural Statistics Service. Retrieved from hitps://www.nass.usda.gov/Charts_and_Maps/Cattle/bcow.php. Accessed 7 July 2016.

Cook R (2014), World Beef Consumption Per Capita (Ranking of' Countries). Beef 2 live. Retrieved from beef2live.com/story-world-beef-consumption-per-capita-rankings-countries-0-111634. Accessed 7 July 2016.

Anonymous (2016c). Total Number of Sheep and Lambs in the U.S. from 2001 to 2015. U.S. Department of Agriculture, National Agricultural Statistics Service. Retrieved from https://www.statista.com/statistics/19440/total-number-of-sheep-and-lambs-in-the-us-since-2001/. Accessed 15 July 2016.

Anonymous (2015). USDA Poultry Production Data. United States Department of Agriculture, National Agricultural Statistics Service. Retrieved from www.usda.gov/documents/nass-poultry-stats-factsheet.pdf. Accessed 21 July 2016.

WILDLIFE AND HUNTING

Introduction

Hunting or producing wildlife species is another way to put food on the table. Our early ancestors survived on wild game and wild food plants before modern agriculture. The dependence on wild animals for protein, vitamins, and minerals is apparent but the death of these animals by some humans is offensive by hunting. Most people don't seem offended by killing of livestock for food or the production of eggs or other animal products from ranches and farms.

The argument today is that humans no longer need to hunt and fish since food can be purchased in the grocery store. However, there are many places around the globe where hunting and fishing are still essential for life and it provides the much needed protein, vitamins, and minerals for survival.

Hunting in America in the early 2000s generated more than $67 billion annually in economic output and produced more than one million jobs in the United States (Anonymous 2002). It generated $25 billion in retail sales, $17 billion in salaries and wages, and employed 575,000 Americans creating sales taxes, state income tax, and federal income tax. Hunting is a positive economic force, steeped in American heritage, family and stewardship for wildlife. Hunting teaches young people the value of wildlife, care of property and companionship.

Hunters are good for the economy by purchasing hunting gear, trucks, boats, coolers, hotel rooms, gasoline, clothing and a host of other items. They also pay a boatload of special excess taxes to the Wildlife Restoration Fund under the Pittman-Robertson legislation for certain hunting equipment and apportions for conservation and education. The average hunter spends nearly $2000 per year and it keeps many communities from economic failure (Anonymous 2002).

In 2011 the U.S. Fish and Wildlife Service reported 13.7 million people aged 16 or older, or about 6% of the U.S. population went hunting and spent $38.8 billion on equipment, licenses, trips, etc. That is $2800 per person per year spent on hunting and it helped create more than 680,000 jobs and helped small businesses in rural communities. Under the Wildlife Restoration Fund passed in 1937 mentioned above, hunters contributed over $1.6 billion for conservation of wildlife and preservation of natural habitats. Overall economic output was reported at $86.9 billion (Anonymous 2013a). Regardless of the numbers, the economic importance of hunting is big business in the U.S.

Game animals need to be managed in each state to preserve their numbers and presence. If there is not much interest in the animal that cannot be harvested, then there may be no interest in supporting them. Game animals need to be appropriately harvested to maintain their habitat and food supply. Overpopulated game animals can destroy the very food supply and habitat that sustains them.

The United States has many areas that wildlife can utilize. However, much of their habitat and food supply has been destroyed by ranching, farms, urban areas, and human activities contrary to their well-being. The positive part is that millions of pounds of wildlife meat (protein) are donated to feed the hungry and poor by hunters each year in addition to their own needs. This also occurs worldwide in remote areas where transport of the meat long distances may not be practical or feasible.

Hunting is also an excellent way to get closer to nature, enjoy family and friends, learn about wildlife and their habitats, learn to handle weapons safely, and learn to camp and become self-sufficient. Hunters will probably always be criticized in this country because some feel we have moved beyond hunting, killing of animals, and disturbing nature. However, our right and desire to hunt and enjoy nature is in our DNA and heritage.

History of Hunting

Hunting is the practice of killing or trapping any animal or pursuing or tracking it.

Hunting wildlife or feral animals is commonly done by humans for food, recreation, or to remove predators dangerous to humans, wildlife, or domestic animals or for trade. Lawful hunting is distinguished from poaching, which is illegally killing, trapping, or capture of hunted species. The species are referred to as game or prey and are usually mammals or birds (Anonymous 2016a).

Hunting can also be for pest control; but hunting has also resulted in the endangerment, extirpation, or extinction of many animals. Pursuing animals without the intent to kill them, such as photography, birdwatching or scientific research, is not hunting.

Human hunting of animals probably reaches thousands of years before domestication of livestock about 11,000 years ago. Hunting strategies resulted in the development of the bow and arrow about 18,000 years ago and domestication of the dog about 15,000 years ago. There is fossil evidence of using spears about 16,200 years ago.

In North America and Eurasia caribou and wild reindeer may have been the species of greatest importance in hunting (Elman 1980; Whisker 1999; Anonymous 2016a).

Hunter-gathering lifestyles still existed in parts of the New World, Sub-Saharan Africa, Siberia, and Australia until the European Age of Discovery. They still exist in some tribal societies, indigenous peoples of the Amazons (Aché), some Central and Southern African (San people), New Guinea (Fayu), the Miabri of Thailand and Laos, Vedde people of Sri Lanka and Hodya of Tanzania (Anonymous 2016a).

Some archaeologists suggest that early hominids and humans were mostly scavengers instead of hunters. However, even after animal domestication became widespread and agriculture was developed, hunting or wildlife was used to supplement human needs. This included meat for protein, bone for implements and weapons, sinew for cordage, fur, feathers and hides for clothing and shelter. Early weapons could be rocks, spears, the atlatl, and bows and arrows.

In modern times hunting is practiced the world over for sport, food supply, and various products of wildlife, such as leather for numerous items. Animal heads and body mounts make decorative displays, especially for the hunter.

The domestication of the dog was used to aid the hunter and led to a symbiotic relationship. Today dogs are human companions and are sometimes used to find, chase, retrieve, and sometimes hunt game. Different breeds do different types of hunting. Waterfowl are commonly hunted using dogs such as the Labrador and Golden Retriever, Chesapeake Bay Retriever, and Brittany Spaniel and other similar breeds. Pit Bull types and others are sometimes used in hunting wild hogs and bears in some states and locations.

In the United States (North America) hunting predates the U.S. by thousands of years and was an important part of pre-Columbian Native American cultures. Native Americans in the U.S. retain some hunting rights and are exempt from some laws as a part of the Indian treaties.

Hunting is primarily regulated by state laws in the U.S. and additional regulations apply to migratory birds and endangered species. Regulations vary from state to state, and some states make a distinction between protected and unprotected (vermin or varmints) species. Protected species require a hunting license and sometimes a tag. Hunting migratory waterfowl requires a duck stamp from the U.S. Fish and Game Service in addition to the appropriate state hunting license. Big game hunting is restricted by a bag limit and a possession limit of a specific animal species an individual can harvest in a single day. A hunting license is required for varmints and predators in some states. These duck stamps, license, and tag fees support conservation, habitat, and wildlife.

Humans, Hunting and Wildlife

Humans evolved and migrated as hunters. Humans are unique among predators because humans can exercise conscious behavior. Humans by self-interest can preserve the habitat and harvest the amount of game needed only for himself, family, and their tribe. The human hunter must consider future needs and habitat of the animals he seeks and is the only predator capable of doing so. In today's world, at least North America, Africa, and other areas, game hunting is highly restricted by law as to the number of animals that can be taken in a given time to preserve the species for future use.

Hunting laws in the early days of the U.S. were not always present as seen in the demise of the bison, elk, deer and loss of the dodo bird by excessive hunting. The same can be said for the demise of rhinoceros for their horns and poaching of other African animals such as elephants for their ivory. In the U.S. bison are making a comeback. The elk numbers in 2009 were over 1 million in North America but were less than 100,000 in the 1890s due to unregulated hunting, grazing competition from cattle, and destruction of habitat (Anonymous 2016b). However, there are an estimated 30 million white-tailed deer in the U.S. today that were nearly extinct over a century ago. Today wildlife managers have the problem of too many deer in some areas or not enough in other areas. Hunters, foresters, farmers, property owners, biologists, wildlife managers, animal right activists, and others interested in effective deer management need to work out acceptable management practices.

Feral pigs are becoming a big problem in the U.S. They do an estimated $1.5 billion in damage to land and crops each year. There are currently more than 5 million wild hogs in the United States and they eat endangered animal and plant species, spread weeds, and many transmit numerous diseases to humans, livestock, and wildlife. Successful control has not been accomplished by trapping and hunting. They produce multiple offspring annually and can escape detection on large areas in heavy cover. They can survive in the wild without human care. They were introduced to North America in the 1500s by Spanish explorers and later were used for hunting and slaughter (see Chapter 9) (Main 2013b). Even today domesticated hogs escape to the wild and interbreed with wild hogs and add to the wild hog numbers.

Management of Game

Although considerable effort is expended in managing and caring for wildlife, at many locations wildlife species depend solely on food and shelter sources found in the wild. In the U.S. and other countries hunters, farmers and wildlife biologists and university personnel improve game species by breeding or selecting for superior animals (good horns, antlers, vigor, disease resistance, good body conformation and size). Adequate nutrition, improved habitat, and predator control is also considered (Weiss 2002).

Each piece of land, ranch, or farm is different and requires different needs depending on soil type, precipitation, location, and game species to accommodate.

The first step in acquired property, whether purchased or leased, is to study maps at the courthouse to get the lay of the land, roads available, and surrounding neighbors. The county Soil and Water Conservation District (SWCD) or Natural Resources Conservation Service (NRCS) office has aerial photographs of all tracts of land. The type of cover, including field crops, forage crops, brush and tree cover in and around the property, will signal the type of habitat. Lease fees as high as $50 per acre can occur, but the national average is about $3 per acre for hunting privileges (Weiss 2002).

By observation and visiting surrounding landowners, one can get a feel for the abundance and type of game animals available. In the U.S. deer hunting would be most common for large game but game birds such as doves, quail, pheasant or ruffed grouse may be present as well as wild turkey. Waterfowl (ducks and geese) may also occur during migratory times of year. Small game, such as rabbits and squirrels, also provide enjoyable sport and delicious table provisions.

With game hunting one may also have to deal with predators (poachers). Don't confront the person but let law enforcement know of your situation. Post your land and avoid personal confrontations and depend on local authorities to solve the problem. Four-legged predators include dogs and cats. If these animals are seen repeatedly chasing game and don't have a collar, these animals need to be dealt with. Wildlife people indicate these free-ranging animals may readily kill young small game animals and birds. Coyotes, when hunting in packs, can decimate deer, and singly small game and nongame species. Bobcats prey heavily on birdlife and small game, but their numbers are usually low and have limited impact. Raccoons and possums focus on nesting sites of birds. Fox prey on small game, waterfowl, and newly-hatched turkey chicks and songbirds. Excessive numbers of these predators can be hunted and trapped.

Deer Management

Planting food plots is a prime way to raise prime deer. As stated in other chapters, the county extension agent, the NRCS, or deer specialists can help in working with land managers.

Some areas devoid of trees first need cover in the form of trees or woody plant cover. Hardwoods are beneficial due to the mast (nuts or seeds) and are succulent browse (leaves, twigs, buds). White oak are favorite trees and predominate in the South in many areas. Other oak trees are good in that they produce mast (acorns). Hardwood species growth can be accelerated by fertilizing (Weiss 2002).

Other areas involve conifer species. They serve as windbreaks around food plots and should be planted in rectangular plantations. Fruit trees of common apple, crabapple, persimmon, pear, plum, and honey locust are used. These trees bear fruit beginning in midsummer through fall (see Chapter 7).

Clover is a food plant deer like, and it is highly nutritious and palatable. It is relatively easy to plant and is a perennial. As stated in Chapters 4 and 5, clovers are commonly grown for hay production, pasture for livestock, or for a green manure crop. There are several clovers or blends of clover that include arrowleaf, crimson, Ladino, Millennium, Osceola, Red, Webfoot, or White Clover. Other wildlife benefit also from these clovers. Lime or fertilizer to maintain these clovers may be necessary. Fall or spring plantings are possible depending upon climate. Seedbed preparation and standard planting and weed control procedures are available.

Alfalfa is another perennial legume deer love and thrive on. It is also eagerly sought by turkeys, rabbits, and geese.

Other food plot plantings commonly include American joint vetch, cow vetch, orchardgrass, ryegrass, triticale and birdsfoot trefoil. These plants provide protein for deer and the seeds are readily sought by turkeys, ducks, doves, and quail (see Chapters 4 and 5).

The reason alfalfa, the clovers, trefoil, orchardgrass, triticale, and others are great for wildlife is because they have been superior livestock forages for a very long time. White clover has been very successful in the no-till areas where logging roads, lowland that flood, dam banks, forest clearings, powerlines, fence rows, and so on can be planted without tillage or seedbed preparation.

Turkeys, geese, doves. quail, ducks, squirrels, songbirds, and many other nongame species benefit from the no-till plants in addition to deer.

Another benefit to good nutrition on specific property is to keep wildlife on your property year-round and not just during hunting season. Soil testing should be done to determine lime and fertilizer needs for the subject property and consider seed blends for best wildlife nutrition. Most seed companies make available to land managers a large variety of seed types that can be planted as stand-alone food plots and customize it as needed.

The most popular varieties for fall and winter include brown-top millet, hybrid pearl millet, sesame, winter wheat, hegari, sunflowers, WGF sorghum, Austrian winter peas, triticale, rye, barley, and winter oats (Weiss 2002).

Late spring and early summer food plots could be WGF sorghum, hegari, alfalfa, soybeans, and summer cowpeas. These plantings can be made on small or large areas.

Corn and soybeans can also be planted. They provide high-carbohydrate and high-energy content. Shrubs, vines, and berries plantings are encouraged. Japanese honeysuckle is a favorite. Native berry plants (blackberry, raspberry, elderberry) are eagerly foraged by deer, gamebirds, and songbirds.

Supplemental feeding of deer is practiced. Shelled corn provides high carbohydrates and energy. Deer chow blends that vary in grain mixture and protein content are available at feed mills. Pelleted wildlife foods consist of several grains, vitamins, and minerals. Supplemental minerals for does and bucks can lead to improved antler development in bucks. Minerals from feed stores are available in powder, granulated, and block form. Phosphorus, calcium, high sodium, zinc, manganese, and copper are important for growth of antlers, teeth, and bone (Weiss 2002).

Turkey Management

If turkeys are on your property and hunting lease, good cover and weed seed and forage plants are required, although during warm weather they depend on insects for some of their diet. Roosting sites are also important. They prefer mature hardwoods with large crowns, but in inclement weather they prefer protection and select conifer species if available. Rye, winter wheat, or oats make good food plots as well as other grasses and legumes. Brush thickets in early spring provide nesting sites for hens. They lay an average of 12 eggs. Of the chicks hatched, some are lost to inclement weather and predation by coyotes, foxes, raccoons, bobcats, dogs, and cats (Weiss 2002).

Upland Game Birds

Upland game birds in North America include doves. quail, pheasants, and ruffed grouse. Their preferred foods are the same as for deer and turkey. They need good cover to hide from predators, especially for nesting sites to rear chicks. Quail and pheasants can be raised in captivity and can be sold to game bird breeders and sportsmen. They can be released into the wild at the poult and adult stages to populate an area. Good nutrition and rearing equipment is required (Weiss 2002).

Small Game

Cottontail rabbits are herbivores and feed almost exclusively on green plants in spring and switch to woody browse during cold weather. Cottontails feed on clover, corn, wheat, soybeans, and many other plants if available. Native foods include weeds such as curly dock, wild carrot, sheep sorrel, the plantains, chicory and clover. Woody plants include sumac, willow, crabapple, maple shoots, and blackberry. They eat leaves, buds, and seeds in spring and summer and twigs, bark, and roots in winter. For optimum denning, hiding, and nesting is to leave brushy fence rows, brush piles, weedy areas and places to hide.

Fox and gray squirrels exist nationwide. They prefer hardwood mast-producing trees if available. Important food sources are grapevines, blackberries, persimmons, wild cherry, hawthorne, and dogwood as do deer, turkey and gamebirds. They prefer holes or cavities in trees for denning. If such holes or cavities are not available, two-foot diameter leaf nests in the forks of any mature tree or den box will provide protection. Unlike rabbits, which obtain water from green forage, squirrels need to drink water periodically. Providing water may keep them close and enhance their numbers (Weiss 2002).

Ducks and Geese

The most important factors for ducks and geese are nesting habitat and food. For geese, aquatic vegetation is relished but so are clovers, alfalfa, and grasses. Geese also feed heavily upon grain when available. Geese seek out nesting habitat in the spring, and the presence of food determines how long they stay on a given piece of property.

Some type of standing water is essential for attracting ducks. It can be a shallow pond or a slow moving stream or river. It can also be a wetland area that periodically floods and has standing water for many weeks of the year. Wood ducks and mallards, however, are more amenable to being kept by landowners than other duck species. Ducks feed on aquatic vegetation and hard mast but also like deer food plots (corn, Japanese millet, and grain sorghum). Wood ducks commonly nest in tree cavities and readily use man-made nesting boxes similar to those made for squirrels.

Controversy of Managed Wildlife

Other game species are managed, especially by federal and state laws, regarding time of hunting and limits on the number of individual species that can be taken. Federal and state programs also provide for improved habitat and feeding programs as required to keep animals alive during drought and winter months; for example, feeding elk after they congregate in the lowlands in some areas of the western U.S. when snow covers the landscape in their summer mountain range. The elk may be a problem on the land in which they congregate causing property damage and sanitary or disease problems as well as excessive consumption of food resources. These animals must either be fed or harvested for meat. Humane treatment of these desperate animals must be considered.

Some managed programs to support wildlife are very good and ensure perpetuation of the species, especially in breeding and selection programs that improve wildlife. However, some individuals indicate that feeding wildlife can make them dependent upon humans and they become unable to survive on their own. Other reasons are animals that lose their fear of humans and become aggressive. Others feel wild animals are unpredictable and dangerous. A fourth reason is the possible exchange of diseases between humans or wildlife through food or close contact.

There are probably many other reasons not to feed wildlife, but support of these animals by humans is needed and timely and limited harvesting, and providing land and natural and supplemental feed is justified. Game and nongame species need to be supported for hunters, sightseers, biologists, and anyone interested in nature. Many species such as the wild turkey, the bald eagle, grizzly bear, wolf, elk, and others have made a comeback in certain areas in North America and without help this would not have occurred.

Many exotics such as the aoudad sheep, axis deer, blackbuck antelope, fallow deer, mouflon, nilgai antelope, sika deer, and wild boar occur in Texas and other states. Many other exotics are being imported and tested, in addition to those indicated, and they require special treatment for monitoring disease, nutrition, and handling (Mungall and Sheffield 1994).

European and North American Game Animals

1. Caribou or Reindeer (*Rangifer tarandus*)
2. Moose or European Elk (*Alces alces*)
3. Red Deer (*Cervus elaphus*)
4. Roe Deer (*Capreolus capreolus*)
5. Fallow Deer (*Dama dama*)
6. Whitetail Deer (*Odocoileus virginianus*)
7. Mule Deer (*Odocoileus hemionus*)
8. Blacktail Deer (*Odocoileus hemionus columbianus and O. hemionus sitkensis*)
9. Pronghorn (*Antilocapra americana*)
10. Mountain goat (*Oreamnos americanus*)
11. Bighorn sheep (*Ovis canadensis*)
12. Dall Sheep (*Ovis dalli*)
13. Mouflon (*Ovis musimon*)
14. Chamois (*Rupicapra rupicapra*)
15. Ibex (*Capra*)
16. Wild Boar (*Sus scrofa*)
17. Brown or Grizzly Bear (*Ursus arctos*)
18. Black Bear (*Ursus americanus*)
19. Polar Bear (*Ursus maritimus*)
20. Cottontail Rabbit (*Sylvilagus*)
21. Swamp Rabbit (*Sylvilagus aquaticus*)
22. Hares (*Lepus*), Whitetail, jackrabbit and others; Brown hare (*Lepus europaeus*)
23. Gray Fox (*Urocyon cinereoargenteus*)
24. Red Fox (*Vulpes vulpes*)
25. Gray Squirrel (*Sciurus carolinensis*)
26. Fox Squirrel (*Sciurus niger*)
27. Raccoon (*Procyon lotor*)

The buffalo (bison) (*Bison bison*) could be added to the long list of animals hunted since limited hunts now occur in North America. Bison provided Indians and early settlers with hair, hide, meat, fuel, bones, and sport. Hunters today use the meat, leather from the hides, head mounts for display, and hunt them for sport.

Caribou

Caribou are the only deer species in which both bulls and cows have horns. The animal's pelt is light in color, thick, an excellent insulator, and waterproof. The animals survive -40°F temperature. Caribou occur in North America, the Lapps, Scandinavia, and into Russia and Siberia. Caribou (reindeer) are one of the most useful deer, yielding fresh milk, hide, meat, and bones. They can be used as draft animals (Palmer and Fowler 1975; Elman 1950).

Moose

Moose are the largest members of the deer family. The European species (Alces *alces*) is found chiefly in Scandinavia where it is known as elk. It can weigh up to 1200 pounds. The North America (Alaska) and Russia (*A. pfizenmayeri*) moose can be up to 1300 pounds with huge horns. Elsewhere in North America the *A. americanus* and *A. andersoni* bulls weigh about 1000 pounds when mature. The Shiras moose (*A. shirasi*) occur in the Rockies and bulls also weigh about 1000 pounds when mature.

All moose need woodlands with bogs and marshes and feed on deciduous coniferous trees. They will graze on aquatic vegetation if needed. They mate in North America in September and October but do not maintain harems. They shed their antlers in winter and grow new ones in spring and summer. Gestation takes eight to nine months and twins are common. Where they occur meat is a significant part of the human food supply. In Sweden nearly 100,000 moose (elk) are taken each year from over a 300,000 estimated population.

Red Deer

The red deer stag (*C. elaphus*) is the preeminent hunting trophy of Europe and has been celebrated for centuries in art, literature, and music. American hunters go to New Zealand and Argentina, South America to hunt the red stag. They also occur westward across Europe from the British Isles to Russia. An average red deer stag of central Europe weighs about 275 to 300 pounds and stands 4 feet tall at the shoulder. If the antlers have up to twelve points they are called royal. The antlers can be very beautiful and impressive in size. The stags fight by butting one another. Red deer are larger in Eastern Europe than anywhere else. They breed towards the end of September being heralded by the roaring stags.

The Wapiti (elk) (*A. canadensis*) belongs to the same species as the red stag but are found from New Mexico and Arizona, U.S.A. to British Columbia, Canada. They attain a weight up to 1000 pounds for males, but the average bull weighs 700 pounds and stands 6 feet 8 inches tall and 9 feet long. The average female is 500 pounds. The male antlers can be 5 feet long with 6 or 7 points per side and are shed each year. They are gray-brown in color. The bulls fight in November and maintain harems of six to eight cows. Calves are born 249 to 262 days after conception. Food is a wide variety of grasses, shrubs, and trees. They provide excellent meat, hides, and head mounts (antlers) for hunters and the general population.

Roe Deer

Other deer species include the roe, fallow, whitetail, mule and blacktail. Roe deer are small compared to the red deer and a mature buck weighs 50 to 65 pounds and stands just over 2 feet at the shoulder. They are found in Europe. East of the Urals and throughout much of Asia, the European roe is replaced by a larger subspecies, the Siberian roe (*C. pygargus*), while another subspecies, the Chinese roe (*C. bedfordi*), is found in China and Korea.

Fallow Deer

Fallow deer originated in Mesopotamia and Valleys of the Tigris and Euphrates. It has been introduced into many other parts of the world, including the U.S. Mature stags weigh about 200 pounds and stand 3 feet tall at the shoulder. They vary widely in color. Only the bucks have palmated antlers like the European elk.

Whitetail Deer

Whitetail bucks range in size up to 200 pounds or more in the Midwest, Pacific West, and Northeast to the tiny Florida Keys. Whitetail (*O. clavium*) are much alike despite being divided by taxonomists into some thirty subspecies of which seventeen are found in the U.S. As indicated earlier there are an estimated 30 million whitetailed deer in the U.S. today and they are prized by hunters and provide abundant meat, hides, antlers, and head mounts. They flourish on the abandoned farms and wildlands and can be as high as fourteen deer or more per square mile.

Mule Deer

Mule deer occur from Mexico to Alaska and in between in the western U.S. There are ten subspecies. They are distinguished by their large ears and bucks develop large antlers. Adult bucks will weigh about 400 pounds. Mule deer are migratory, spending summer at high altitudes and descend to winter pastures starting in September. Breeding takes place toward the end of October. The bucks do not form harems nor do they fight much. The antlers and meat are prized by the American hunter.

Blacktail Deer

Blacktail deer inhabit the Pacific coastal forests of North America. Their tails are black and they have a white underside. The blacktail is two subspecies of mule deer, the Columbian and Sika blacktail, but differ from the mule deer in a number of respects. They breed in mid-September in the south of their range to November in the North. Gestation is about seven months.

Pronghorn

A mature buck weighs up to 145 pounds and up to 41 inches tall at the shoulder. Both sexes are horned (permanent and not shed). The horn can grow to 20 inches in length but about 12 inches is average. They consist of an inner core and outer sheath, the latter sheds annually. Female horns rarely exceed a length of 3 inches. Pronghorns are light to dark brown with white underparts. If alarmed, the long white hair on its rump patch erects. Pronghorns are found in the western part of North America. They are a favorite game animal for open plains hunting and provide meat, trophy heads, and hides. Their main food is grass, herbs, mosses, and lichens.

Wild Sheep and Goats

I am going to group the wild sheep and goats together since they are more specialty hunts and less frequent than other big game even though their meat, hides, and heads are used. These hunts are for the strong adventurer since most hunts occur at high altitude in rugged country.

The mouflon is a wild sheep established in North America. It was originally introduced to Europe from Sardinia and Corsica. The chamois and ibex are European wild goats. The chamois male weighs about 75 pounds and the ibex is bigger at 230 to 240 pounds for mature males. Alpine and Spanish ibex exist in Europe.

The mountain goat is an antelope found only in North America in the mountains of Alaska, the Yukon, and the Northwest Territories south through British Columbia and Alberta into Washington, Montana, Idaho, Wyoming, and South Dakota. Males or billies weigh 200 to 300 pounds and the females or nannies are slightly smaller. Both sexes are horned (spike-like) with white coats and long guard hairs. Their hooves have a spongy portion that grips like a suction cup on steep rocky surfaces. Nannies and kids stay together in herds while billies are solitary. Mating starts in November. They feed mostly on grass in the high mountain meadows. Billies fight during mating season. Kids are born in April to June. The goats seek lower elevations in winter.

Bighorn sheep have spectacular, heavy, curved horns and inhabit high mountain country. The bighorn is North America's most coveted trophy. Two classes occur: the Rocky Mountain bighorn and the desert bighorn for purposes of scoring trophies. The range extends from British Columbia and Alberta south through the western U.S. mountains into northern Mexico. The arid hills of Nevada, California, New Mexico, Arizona and northern Mexico comprise the habitat of the desert bighorn. Both sexes are horned. The males can weigh up to 300 pounds. Bighorns are generally brown with pale snouts, rump patches, and bellies. Mountain bighorns are usually dark brown. Mating season is October and November. Bachelor herds of males occur after the mating season. They go to lower country to overwinter.

Dall sheep occur in Alaska, the Yukon, and the Northwest Territories while its subspecies, the stone sheep, is found in British Columbia. The Dall is white with thinner horns than the bighorn. Their horns flare out at their tips. The stone sheep is dark gray to almost black and its horns are heavier than those of the Dall but have the same shape and structure. Both Dall and stone are similar to the bighorn in life cycle.

Wild Hog

Columbus on his second voyage in 1493 to America brought swine that populated the West Indies and later by others who brought swine to North America (see Chapter 9). Wild boar were transplanted to North America in the late 1890s and early 1900s (Anonymous 1980). Some of these wild pigs escaped and spread, often interbreeding with stray domestic pigs or their feral wild boar relatives. Coloration and body conformation is often different in North America than their European cousins.

Wild boar are still found in Europe eastward from West Germany and, like the red deer, the wild boar is generally bigger eastward. The average weight of males in Germany is about 200 pounds and almost 400 pounds in Turkey.

Wild pigs are omnivores and cause great damage to hay and field crops in Europe and the U.S. where they occur. They are hunted and trapped relentlessly but they continue to persist and cause problems.

Wild boars are armed with large tusks as they mature. They protrude to 8 inches per side of the snout and are very sharp and dangerous. Weights of wild hogs vary from 200 to 350 pounds in the U.S. The wild pig is a great source of meat and sport for the hunter.

Bears

In Europe the brown bear exists in small numbers in the Pyrenees, the Alps, the Balkans, and northern Scandinavia. It is also found in Russia and Asia. In North America their range extends from Idaho, Montana, and Wyoming north through British Columbia and western Alberta and the Northwest Territories, the Yukon, and the interior of Alaska. The Alaskan Brown (coastal) bear ranges from southern Alaska, including the Kodiak Islands, south to British Columbia. The coastal bear is bigger than the grizzly because it eats more fat and protein. Female coastal bears weigh between 500 to 800 pounds. Mature males weigh 800 to 1200 pounds. Brown bears are omnivores.

Black bears are much smaller than the brown bear and weigh 200 to 600 pounds and are omnivorous. Bear meat can be eaten if properly prepared. Since bears are omnivorous, trichinosis can be a problem; and the meat, like pork, needs to be well cooked. Some say bear meat is good when properly prepared but others reject it.

Rabbits and Hares

Cottontail rabbits are popular game animals in North America and are found everywhere in the United States. There are mountain, desert, New England and eastern cottontails, and they are a good source of meat and sport.

Swamp rabbits can weigh up to 6 pounds and are larger than cottontails. They are found in the southern and central U.S. in swamps and marshes. They provide good sport and meat.

Hares are found in North America, such as the white-tailed jackrabbit and black-tailed jackrabbit. The arctic hare, tundra hare, snowshoe and varying hare are also different species. The brown hare is the hunted hare in Europe. The Alpine and blue hare also occur there. The brown hare has also been transplanted to the U.S. Hares are important food for foxes, hawks and eagles and other predators.

Squirrel

The gray squirrel is one of the most sought-after game animals in North America. It ranges throughout the U.S. and southern Canada. It averages about 1 pound in weight. They inhabit wooded areas and eat nuts, acorns, and corn. The fox squirrel occurs in the same parts of the U.S. and Canada as the gray squirrel. It is slightly bigger than the gray squirrel (Anonymous 1980).

Exotics in the United States

There are also many exotic animals in the United States (Texas) on our pastures, rangelands, and forests. Some newly established exotics include the Aoudad, axis deer, blackbuck antelope, fallow deer, mouflon and mouflon crosses, nilgai antelope, sika deer, and wild boar. Fallow deer, mouflon, and wild boar have already been mentioned (Mungall and Sheffield 1994).

There are also many other exotics such as the addax, barasingha, eland, ibex (already mentioned), greater kudu, oryx and gemsbok, red deer (already mentioned), sable antelope, sambar, sitatunga, waterbuck and zebra and others that are being studied for adaptation and hunting. All these exotics provide meat and trophies for the hunter and sport. They sometimes require special care, nutrition, and habitat similar to their native origin. Some of the animal associations are against bringing in these exotics because of disease problems, competition for habitat with domesticated animals, and competition in the meat market.

Upland Fowl

The modern turkey is a descendant of the wild turkeys native to North and Central America. Domestic ducks and geese were also once wild but can be raised for meat, eggs, and plumage like chickens (see Chapter 9). However, there are numerous wild game birds that are entirely wild and provide meat and sport for the hunter and wildlife observer. They include the woodcock, snipe, quail, turkey, dove, pigeon, grouse, ptarmigan, prairie chicken, ring-necked pheasant, and partridge.

Woodcock

Popular hunting species include the American Woodcock (*Philohela minor*) which is essentially the same species in appearance as the European woodcock even though they have different Latin names. They are a dark brown, plump bird of wet and swampy woods. They feed on insects and some plants and are a small game bird.

Bobwhite Quail

The bobwhite quail (*Colinus virginianus*) is a classic American upland hunting bird. It feeds in grain fields and fence rows. They are small, generally 8.5 to 10.5 inches tall, and are brownish with light and dark reddish tints in some feathers. They gather in groups called bevies of 12 up to 30. There are other quail species including scaled, Gambel's, California, mountain, means, and crested. They are all hunted.

Wild Turkey

The wild turkey (*Meleagris gallopavo*) includes at least half a dozen subspecies which itself looks similar to the domesticated turkey. The males are much larger than the females and have more intensely red colored wattles. All males have beards and some females have beards. The bird feeds on acorns, walnuts, hazelnuts, beechmast, etc. They roost in trees and are difficult to approach and hunt.

Doves

Mourning dove (*Zenaidura macroura*) and white-wing dove (*Z. asiatica*) nest in Canada, the United States, and Mexico. They flock together often after harvest in grain fields. The white-winged doves have conspicuous white patches on the wings. Mourning and white-winged doves are similar in size (11 to 12 inches) but the white-winged dove has an owlish call compared to the coo of the mourning dove and favors a more arid climate. They are hunted on agricultural lands and the white-winged dove is also hunted in Central America.

Ptarmigan

Ptarmigan (*Lagopus*) occurs in the Arctic tundra of North America, Greenland, Iceland, the Scandinavian mountains, and Siberia. Their southern limits are the northern U.S. They are white with black markings in winter that vary from subspecies to subspecies and are generally brown with white wings in summer. Ptarmigans are similar in size to the partridge, or slightly bigger. They provide food for humans and food for furbearers in their range.

Grouse

There are several species of grouse including the red, ruffed, spruce, sharp-tailed, blue, sage, black, greater prairie chicken, lesser prairie chicken, hazel grouse and capercaillie. The ruffed, spruce and sharp-tailed, blue and sage grouse are found in the U.S. and some areas are hunted. The lesser prairie chicken is most numerous in Kansas and Oklahoma and both states have hunting seasons for them. Most of the grouse are relatively small. The capercaillie is the largest European grouse. The cock weighs between 8 to 12 pounds. The black grouse and hazel grouse of Europe are similar in size with males of about 2.6 pounds and females at 1.8 pounds. The ruffed grouse is 19 inches in length and weighs 29 oz. The sage grouse is the largest North American grouse with the male at 6 to 8 pounds and the female 3 to 5 pounds in weight (Palmer and Fowler 1975; Anonymous 1980).

Ring-necked Pheasant

Ring-necked pheasant (*Phasianus colchicus*) cock is a long-tailed and brightly-colored pheasant with a white ring around its neck while the hen is smaller, brownish with a long tail. This pheasant is indigenous to Asia Minor and was introduced into North America in the 1880s and is hunted and widely established in North America. Males are 3 feet in length and hens 20 inches in length. They can weigh up to 4.5 pounds when mature (Palmer and Fowler 1975; Anonymous 1980). The ring-necked pheasant and upland birds sometimes are raised like poultry and released into the wild when ready.

Partridge

The Hungarian partridge (*Perdix perdix*) was introduced into North America from Hungary. There are now large populations in the Pacific Northwest and the Dakotas. They are gray underneath with a large spot of reddish brown in adults. They are 14 inches in length and weigh 1.5 oz. Males are slightly heavier. They appear intermediate between grouse and the bobwhite in size. They love stubble and grain fields for both roosting and feeding. Covies number up to about twenty birds and are fun to hunt.

The Hungarian partridge can suddenly surprise the hunter by a covey suddenly flushing up nearby. Their wings make a loud buzzing noise in flight and they usually land out of the hunter's reach. The author has experienced this many times on the Camas Prairie in northern Idaho.

Other partridge species include the chukar partridge (*Alectoris graeca* and *A. chukar*) introduced into the U.S. from Europe and inhabit the Pacific Northwest. The chukar are only about 13 inches long and get their name from the sound of its call. The red-legged partridge (*Alectoris rufa)* is native to southwestern Europe and is identified by its red bill and legs, heavily barred flanks, long white stripe above the eye, and pale orange-chestnut-colored face. It also has a conspicuous dark brown horseshoe mark on its lower breast. It is a fast runner and flier and is a challenge for the hunter (Anonymous 1980).

Waterfowl

Waterfowl are made up of geese and ducks. The number of species is numerous. We discussed the domestic rearing of geese and ducks in Chapter 9. However, geese found in the wild live in a different habitat and world.

Geese

Canada Geese

The Canada goose or honker (*Branta canadensis)* is 3.5 feet long with a wing spread of 5.5 feet and weighs 18 pounds. The female is smaller. The bill and feet are black. Their wings and backs are a dark gray-brown and their breasts, flanks, and undertail covers are white. They fly in a V-formation. They nest in the northern part of the continent and as far south as the northern U.S. Most winter in the mid-U.S. to the Gulf of Mexico. They can become troublesome in corn, wheat, and soybean fields. They are a favorite hunting species (Palmer and Fowler 1975; Anonymous 1980).

Snow Geese

Snow geese (*Chen hyperforea)* are 31 inches maximum with a wing length of 16 inches. They weigh 2 to 3 pounds and are white with black wing tips. The bill is red with black edges and dark red feet. The females are smaller. The lesser snow goose (*C. caerulescens*) winters mainly in the Mississippi Delta, Gulf Coast, and scattered areas of Mexico and California. The greater goose (*C. atlantica*) winters along the Atlantic coast. The lesser and greater snow geese are subspecies and they nest in the extreme north of the American continent. These geese like grain fields after harvest but can cause problems in some areas of central Canada before fields are harvested. From personal experience, snow geese and Canada geese can be harvested for their meat.

There are many more species of geese including the Ross, Emperor, white-fronted, Magellan, ashy-headed, graylag, bean, pink-footed, barnacle, and brant (Anonymous 1980).

Ducks

The mallard (*Anas platyrhynchos*) is a common bird of the entire northern hemisphere. The brightly-colored drake has a green head and neck, gray-brown body feathers, and typical curled tail. The female has no collar and with a brown head and neck, wing bar is purple. The length of the drake can be 28 inches, wingspread to 40 inches, and weighs up to 3.75 pounds. The female is somewhat smaller than the drake. The birds are abundant and provide excellent meat. They can be raised in captivity.

Other ducks include the shoveler, gadwall, pentail, black, wood, fulvous tree duck, canvasback, redhead, greater scaup, lesser scaup, seaducks, common teal, green-winged teal, blue-winged teal, cinnamon teal, baikal teal, speckled teal, American widgeon and European widgeon.

African Game Animals

Some of the animals hunted for meat, hides, and trophies will be listed for Africa, Asia, and South America for completeness of this chapter. The African Plains game are listed as follows:

1. African Buffalo (*Syncerus caffer*)
2. Bongo (*Boocerus eurycerus*)
3. Buckbuck (*Tragelaphus scriptus*)
4. Eland (*Taurotragus*)
5. Greater Kudu (*Strepsiceros strepsiceros*)
6. Lesser Kudu (*Strepsiceros imberbis*)
7. Nyala (*Tragelaphus angasii*)
8. Sable Antelope (*Hippotragus niger*)
9. Roan Antelope (*Hippotragus equinus*)
10. Lechwe (*Kobus leche*)
11. Reedbuck (*Redunca*)
12. Hartebeest (*Alcelaphus*)
13. Wildebeest (Connochaetes)
14. Grant's Gazelle (*Gazella granti*)
15. Impala (*Aepyceros melampus*)
16. Waterbuck (*Kobus ellipsiprymnus* and *K. defassa*)
17. Warthog (*Phacochoerus aethiopicus*)
18. Bush pig (*Potamochoerus porcus*)
19. Addax (*Addax nasomaculatus*)
20. Scimitar-horned Oryx (*Oryx algazel*)
21. Gemsbok (*Oryx gazella*)
22. Hippopotamus (*Hippopotamus amphibius*).
23. Crocodile (*Crocodylus niloticus*)
24. Sitatunga (*Tragelaphus spekii*)

Asian Game Animals

1. Tahr (*Hemitragus jemlahicus*)
2. Blue Sheep (*Pseudois nayaur*)
3. Takin (*Budorcas taxicolor*)
4. Serow (*Capricornis*)
5. Goral (*Naemorhedus goral*)
6. Tibetan Black Bear (*Selenarctos thibetanus*)
7. Sloth Bear (Melursus ursinus)
8. Malayan Sun Bear (*Helarctos malayanus*)
9. Markhor (*Capra falconeri*)
10. Siberian Ibex (*Capra ibex sibirica*)
11. Barking Deer (*Muntiacus muntjac*)
12. Hog Deer (*Axis porcinus*)
13. Chital Deer (*Axis axis*)
14. Nilgai or Blue Bull (*Boselaphus tragocamelus*)
15. Blackbuck Antelope (*Antilope cervicapra*)

South American Game Animals

1. Marsh deer (Blastocerus dichotomus)
2. Guemal (*Hippocamelus*)
3. Savanna deer (Odocoileus gymnotis)
4. Peccary (*Tayassu*)
5. Tapir (*Tapirus*)

A number of game species have been introduced into South America or, more accurately, Argentina including the red deer, fallow deer, chital deer, blackbuck, and wild boar. Many of the same animals have been introduced into New Zealand as well as tahr, chamois, and others. Some of the cats and non-meat animals are not included in this chapter for the U.S. and world.

Summary

Hunting or producing wildlife species is another way to provide food. Our ancestors survived on wild game and food plants for protein, vitamins, and minerals. The argument with modern humans is that hunting is no longer needed since food can be purchased in the local grocery store. However, there are many places on the globe where hunting and fishing are still essential for life.

In the U.S. recent economic evaluation suggests that over 86 billion dollars are annually generated by the hunting industry in terms of jobs created, equipment, lodging, food, and hunting licenses purchased. From these purchases considerable tax funds are used for conservation and education for wildlife and habitat.

In 2011 the average hunter in the U.S. spent $2800 per year on hunting by nearly 14 million hunters age 16 or older. These expenditures helped small businesses in rural communities and the overall economy.

Individuals and government spent considerable funds and resources in management of game animals and game birds. Game animals need space and a good habitat to prosper; but they also need to be harvested to maintain their population, habitat, and food supply.

In the United States we have many areas that wildlife can utilize, however much of their habitat and food supply has been destroyed by ranching, farming, urbanization, and other human activities contrary to their well-being. The positive aspect is that millions of pounds of deer, elk, pig, and game birds are donated to feed the hungry and poor each year in addition to the hunters' own needs. The same occurs worldwide in remote areas where transport of the meat long distances may not be practical and is given to the local people.

Hunting is also a way to get closer to nature, enjoy family and friends, learn about wildlife and their habitat. The desire to hunt and be in nature is in our DNA and heritage.

Human hunting of animals reaches back thousands of years before domestication of livestock about 11,000 years ago. Hunting strategies resulted in the development of the bow and arrow about 18,000 years ago and domestication of the dog about 15,000 years ago.

In North America and Eurasia, caribou or wild reindeer may have been the species of greatest importance in hunting and domestication.

In modern times, hunting is practiced the world over for sport, food supply, and various products. The domestication of the dog was used to aid the hunter and led to a symbiotic relationship. Today dogs are human companions and sometimes are used to chase, retrieve, and hunt game animals. Different breeds do different things like retrieving upland birds, waterfowl, or various game animals.

In the U.S. hunting predates the U.S. by thousands of years and was an important part of the pre-Columbian Native American culture. Native Americans in the U.S. retain some hunting rights and are exempt from some laws as part of the Indian treaties.

Humans as hunters are unique predators because they can exercise conscious behavior and by self-interest can preserve the habitat and game harvest. The hunter must consider future needs and habitat of the animals he seeks. In today's hunting world, hunters are highly restricted by law to the numbers and types of animals that can be taken to preserve the species for future use.

In the U.S. the demise of the buffalo (bison), elk, and deer occurred due to excessive hunting in the 1800s. Bison are making a comeback and elk numbers are over 1 million from 100,000 in the 1890s. Deer were nearly extinct a century ago in the U.S. but now there are an estimated 30 million.

Feral pigs are becoming a big problem in the U.S. They do an estimated 1.5 billion dollars in damage to land and crops each year. There are currently more than 5 million wild hogs in the U.S. Trapping and hunting reduce numbers, but they remain a problem since they can hide and escape in heavy cover on large land areas. However, they provide sport hunting and much-needed protein for the needy and hungry.

Considerable effort and cost is expended in managing and caring for wildlife in the U.S. and other countries. Hunters, farmers, and wildlife biologists work to improve deer quality by breeding programs and by improved nutrition and habitat. Each piece of land, ranch, or farm is different and requires different needs depending on soil type, precipitation, location, and game species. Turkey, geese, doves, quail, ducks, squirrels, songbirds, and many other non-game species benefit from good deer management.

Native animals hunted in Europe and North America are discussed as well as exotics in the United States. Upland fowl and waterfowl are described. A list of African, Asian, and South American game animals are listed that provide meat, hide, mounts, and sport.

Chapter 10. Literature Cited

Whisker JB (1999). *The Right to Hunt.* Merril Press, Bellevue, WA. 207 p.

Elman R (1980). *The Complete Book of Hunting.* Abbeville Press, New York, NY. 320 p

Weiss J (2002). *Planting Food Plots for Deer and Other Wildlife.* Woods N Water, Inc. Bellvale, NY. 164 p.

Holechek JL, Pieper RD, Herbel CH (2002). *Range Management: Principles and Practices.* Fourth Edition. Pearson Custom Publishing, Boston, MA. 571 p.

Anonymous (2002). Economic Importance of Hunting in America. International Association of Fish and Wildlife Agencies. Retrieved from www.fishwildlife.org/files/Hunting_Economic_Impact.pdf. Accessed 25 July 2016.

Anonymous (2013a). Hunting in America: An Economic Force for Conservation. National Shooting Sports Foundation. Retrieved from www.nssf.org/..huntinginamerica_economicforceforconservation.pdf. Accessed 25 July 2016.

Anonymous (2016a). Hunting. Wikipedia. The Free Encyclopedia. Retrieved from https://en.wikipedia.org/wiki/Hunting#United States. Accessed 26 July 2016.

Anonymous (2016b). Elk Numbers Across 6 States. 2016 goHunt, LLC. Retrieved from www.gohunt.com/read/elk-numbers-across-6-states. Accessed 28 July 2016.

Rooney TP (2010). What Do We Do With Too Many White-Tailed Deer? Action Bioscience. Retrieved from www.actionbioscience.org/biodiversity/rooney.html. Accessed 28 July 2016.

Main D (2013b). Feral Pigs Going Hog-Wild in U.S. Livescience. Retrieved from www.livescience.com/28560-feral-pigs-running-wild.html. Accessed 28 July 2016.

Mungall EC, Sheffield WJ (1994). *Exotics on the Range: The Texas Example*. Texas A&M University Press. College Station, TX. 265 p.

Anonymous (1980). Game. Chapters 1-6, Part II. Pages 26-116 in Elman R ed. The Complete Book of Hunting. Abbeville Press, New York, NY.

Palmer EL, Fowler HS (1975). *Fieldbook of Natural History. 2nd Edition*. McGraw-Hill Book Company, New York, NY. 779 p.

FISH FARMING

Introduction

Fish farming or pisciculture is raising fish commercially, usually for food. A facility that releases juvenile fish into wild waters for recreation fishing or to supplement the species in nature is generally referred to as a fish hatchery. The most important fish species worldwide using fish farming are carp, tilapia, salmon, and catfish (Anonymous 2016).

According to Wikipedia, the free encyclopedia, the demand for fish has caused widespread overfishing in the wild and by 2016 more than half of the seafood is produced by aquaculture. China is a big producer of farmed fish. The 2008 global returns for fish farming recorded by FOA totaled nearly 40 million metric tons worth about $60 billion in the U.S.

Requirements for Recreation

Waters et al. (1978) outlines the requirements for growing fish. It can be done for personal recreational use or to sell for profit. It requires clean water, a suitable site for a pond or tank(s), planning and a great deal of knowledge, especially for a commercial operation.

One must have the land (soil, water, topography, and other resources) to be successful. The soil must hold water and have enough spring flow, well water, or runoff to fill the pond in a year or less and replenish water when lost from seepage or evaporation. Water quality must also be adequate for the species of fish grown. It will be either cold or warm water depending on species grown. Warm water ponds managed for bass and bluegill for recreation should be an acre in size or larger. Trout ponds should be 1/3 acre at minimum. If the pond is covered in ice for a month or more, winter fish kills are frequent and should be at least 20 feet in water depth.

In the South shallower water depth of 3 to 4 feet is adequate for fish growth.

Spillways to prevent floodwater from going over the dam should be designed to keep water flow shallow to prevent loss of fish swimming out or in.

Stocking Fish

Stocking fish for recreation and commercial purposes for warm water greater than 80°F (average summer temperature) include blue catfish, bluegill, buffalo fish, carp, channel catfish, crappie, fathead minnow, golden shiner, goldfish, hybrid sunfish, largemouth bass, red ear sunfish and white catfish. Stocking for average summer water temperature of less than 70°F include brook trout, rainbow trout, northern pike, and muskellunge. Stocking rates depend on fish species, water fertility, and supplemental feed and fertilizer used. In warm water ponds (Alabama) 50 bass and 500 bluegill are stocked per surface acre on average, whereas ponds maintained at high fertility with commercial fertilizer are stocked at 100 to 200 bass and 1000 to 1500 bluegill. In cold water areas, average ponds support 500 half-pound trout per surface acre annually, whereas supplemental feeding of the same pond will produce 2000 half-pound trout per year (Waters et al. 1978).

Management of Fish

Intensive management can produce exceptional fish production. However, the cost of fertilization, feeding, chemical control of aquatic weeds, supplemental stocking, and other needs may be more costly than most people are willing to pay for a few hours of fishing. One must decide on the level of management that fits your time and pocketbook.

Fertilization effectiveness is generally reduced in northern states and may not be recommended. In the South a mineral fertilizer is best for fish ponds. For bass and bluegill, 8 pounds of nitrogen, 8 pounds of phosphate, and 2 pounds of potash per surface acre (100 pounds of 8-8-2 fertilizer) is commonly recommended for each application. When highly managed, fertilizer can be applied each month. Feeding can also be increased to harvest more fish as well as weed control and other good management practices (Waters et al. 1978).

Commercial Operations

Commercial operations involve raising fish for human food, bait, stocking, and fee fishing. Additional enterprises produce a brood fish, eggs, and ornamental fish.

Commercial operations require a larger investment and more knowledge generally than raising fish for your own recreation and use. First a dependable, ample and good-quality water source is essential. Well water, springs, streams and runoff ponds are suitable. Most commercial trout are raised in raceways (a channel with a current of water) that flow all year from springs, wells, or streams. The number of trout raised each year depends on water volume. A minimum flow of 450 gallons per minute is required (Waters et al. 1978). Well water is the best source because it avoids unwanted organisms, floods, pollutants, and muddiness. Fresh water should replace pond water from evaporation and oxygen supply every week.

Springs are a good source of water but may contain undesirable fish. Springs must supply adequate water during the dry season. Aeration may be required if oxygen levels become inadequate. Undesirable fish can be removed with an appropriate fish toxicant or by filtering the water. Undesirable fish can also come from pond water, streams, bayous, canals, or other water sources and steps must be taken to keep them out. The Natural Resource Conservation Service (NRCS) technicians can be helpful in planning the fish raising operation.

Commercially Managing Water and Fish

As stated, a dependable year-round well is an excellent source of water. Well water frequently contains little dissolved oxygen and large amounts of carbon dioxide and nitrogen, a mixture deadly to fish. This can be remedied by spraying water into the air or splashing it over a hard surface before it enters the water where fish are raised (Waters et al. 1978).

Prevent unwanted fish from entering your operation rather than attempting to remove them after they enter. Use a dependable year-round water source free of fish if possible; and if surface water is used, filter it before it enters your operation.

To prevent disease and parasites, use clean water. Bring in only fish free of disease and parasites and use clean feed free of disease and parasites.

Oxygen deficiency is a major cause of fish kills, especially in high-stocked operations. Stock only recommended numbers and species of fish, use high-quality water at the recommended depth, and feed recommended amounts of feed.

If fertilizer is needed, use only inorganic fertilizer and prevent runoff from feedlots, poultry operations, pig pens, and dairy barns into the fish operation.

Fish may be harvested during any season, but it is usually best during cold weather. Facilities for dressing and storing a large volume of fish may be required depending on the size of the harvest. Restocking is another key part of the management. Some need a complete restocking every year. Some operations need only periodic restocking depending on the operation.

Other management details include care of the watershed, protection from livestock wastes, treatment of disease and parasites, weed control, and control of wildlife predators such as beaver and muskrat.

Marketing and Costs

Having a viable market available is paramount to your fish-raising enterprise and should be checked out before investing.

Costs to get started can be very high including land, ponds, raceways, wells, drainage pipes, fences, roadways, boats, handling facilities, etc. Yearly maintenance is another cost which includes stocking, pumping water, feed, fertilizers, chemicals, labor, taxes, and other costs.

Your purpose is to produce a high-quality product for the lowest possible cost. Market demand and a contract to furnish fish produce to a local market is desired. Being near a population center is helpful. High-volume operations should be equipped to process and store large volumes of fish. Assistance and information is available from the Cooperative Extension Service, Natural Resources Conservation Service, state game and fish agency and the U.S. Fish and Wildlife Service. The local fish biologist and wildlife specialists and other commercial producers should be very helpful.

Methods of Fish Farming

Intensive Aquaculture

The method of fish farming described so far in this chapter is called intensive aquaculture. It requires, as described, sufficient oxygen to prevent death of fish and a high level of protein (up to 60%) because of their high conversion efficiency (FCR) (kg of feed per kg of animal produced). Salmon have an FCR of 1.1 kg of feed per kg of salmon produced versus an FCR for chicken at 2.5 kg per kg of chicken produced. Fish compared to chicken and other warm-blooded animals do not require as much fat and carbohydrate in the diet for energy.

In very high intensity recirculating aquaculture systems (RAS) there are strict controls over production parameters for high valued species. Therefore, RAS is only economical for high-value products like brood stock for egg production, fingerlings for net pen aquaculture operations, sturgeon production, and special niche markets (Anonymous 2016).

Extensive Aquaculture

The other method of raising fish is by extensive aquaculture where fish are limited by available food such as zooplankton, pelagic algae or benthic animals such as crustaceans or mollusks. Tilapia can feed on phytoplankton.

Algal blooms can be a problem because optimal conditions for fish also favor algae growth. The algae can multiply at a very rapid rate and deplete the nutrients, and their dead biomass can deplete the oxygen in the water by blocking out sunlight. Depleted nutrients, oxygen, and addition of pollutant chemicals from decaying algae can lead to massive fish kills. To remedy these problems, different fish species are used to occupy different places in the pond ecosystem. For example, a filter algae feeder is used for tilapia, a benthic feeder for carp or catfish, and zooplankton feeders for various carp or submerged weed feeders such as grass carp (Anonymous 2016).

Types of Fish Farms

Cage System

Fish cages are placed in lakes, bayous, ponds, rivers, and oceans to contain and protect fish until harvest. Fish are stocked in cages, fed, and harvested when market size. Copper alloys are important netting materials for cages since they destroy bacteria, viruses, fungi, algae and other microbes and provide a healthy environment.

Irrigation Ditch or Pond System

The basic requirement is to have a ditch or pond that retains water with an aboveground irrigation system. Fish can be fed commercial fish food and their waste can be used to fertilize the fields. Good water quality and preventing eutrophication to maintain high oxygen levels is essential.

Composite Fish Culture

This is a fish culture system developed in India where five or six fish species are used in the same pond. The species are selected based on different food habits so they don't compete. The system includes catla and silver carp (surface feeders), rohu (a column feeder), and mrigal and common carp (bottom feeders). Other fish feed on the common carp excreta so fish production can be as high as 6000 kg per hectare per year (14,820 pounds per acre per year).

Integrated Recycling System

Basically, large plastic fish tanks are placed in a greenhouse. A hydroponic bed is placed nearby. Tilapia are grown in the tanks and feed on algae which occur naturally when fertilized. The tank water is slowly circulated in the hydroponic beds and the tilapia waste feeds commercial crop plants. Cultured microorganisms in the hydroponic bed convert ammonia to nitrates and the crop plants are fertilized by the nitrates and phosphates. Other wastes are removed from the hydroponic media and aerated in the same process.

Classic Fry Farming

Classic fry farming is a flow-through system where trout and other sport fish are often raised from eggs to fry or fingerlings and then transported to streams or ponds for release. They are raised in long shallow concrete tanks fed with fresh stream water. They are fed commercial fish pellets.

Additional Considerations

Fish Food

Tilapia, carp, catfish, and many others require no meat or fish products in their feed. Carnivorous fish, like salmon species, depend on fish feed with a portion from a wild catch such as anchovies, menhadens, etc. Vegetable-derived proteins are successfully used in fish feeds for carnivorous fish, but vegetable-derived oils have not been successfully incorporated into the diet of carnivores (Anonymous 2016).

Fish Concentration

Most successful aquaculture species are schooling species and do not have social problems at high density. In nature fish tend to aggregate into large schools. The stocking rate was discussed in the stocking fish section above. Stocking rate depends on fish species, water fertility, feed, and fertilizer used. Large fish concentrations contribute to habitat destruction. The high concentration of feces need to be flushed out and disposed of regularly. If contaminated, fish farms need to be moved to uncontaminated areas.

Disease and Parasites

Sea lice can cause deadly infections in both farm grown and wild salmon, particularly in highly populated concentrations. Some managers use antibiotic drugs that are controversial because of possible escape of these drugs to the environment or possible exposure in human food products.

Disease problems constitute the largest single cause of economic loss in U.S. aquaculture. In 1988 channel catfish producers lost over 100 million fish worth nearly $11 million. Trout producers reported 1988 losses of 20 million fish worth over $2.5 million. Bacterial infections constitute the most important source of disease problems in all the various types of production. Fungal disease constitutes the second most important losses, especially in the crustaceans and salmon. External protozoan parasites cause a loss of large numbers of fry and fingerling fish and cause epizootics in young shellfish. The number of therapeutants approved by the FDA is limited (Meyer 1991).

Slaughter Methods

Electric or percussive stunning methods are being used for more humane slaughter of fish (Anonymous 2016).

Summary

Fish farming or pisciculture is raising fish commercially used for food. The facility that releases juvenile fish into the wild waters for recreation fishing or to supplement a species in nature is generally referred to as a fish hatchery. The most important fish species worldwide used for fish farming are carp, tilapia, salmon, and catfish. In 2016 more than half of the seafood or fish was produced by aquaculture.

Growing fish can be done for personal recreation use or for profit. It requires clean water, a suitable site for ponds or tanks, planning, and a great deal of knowledge, especially for commercial operation. Stocking fish for recreation and commercial purposes for warm water greater than 80° (average summer temperature) include blue catfish, bluegill, buffalofish, carp, channel catfish, crappie, fathead minnow, golden shiner, goldfish, hybrid sunfish, largemouth bass, redear sunfish, or white catfish.

Stocking for average summer water temperature of less than 70°F includes brook trout, rainbow trout, northern pike, and muskellunge. Stocking rates depend on fish species, water fertility, and supplemental feed and fertilizer used. In warm water ponds 50 bass and 500 bluegill are stocked per surface acre on average, whereas ponds of high fertility can be stocked at 100 to 200 bass and 1000 to 1500 bluegill. In cold water areas average ponds support 500 half-pound trout per surface acre annually but can support 2000 half-pound trout per acre with supplemental feed.

Intensive aquaculture requires sufficient oxygen to prevent death of fish and a high-protein feed with good management. Extensive aquaculture is raising fish on food available in their environment.

The types of fish farms include cage systems where fish cages are placed in lakes, ponds, rivers, or oceans to contain and protect them until harvest. Fish are stocked, fed, and harvested when market size. The second type is an irrigation ditch or pond system that retains water with an aboveground irrigation system. Fish can be fed commercial fish food and their waste can be used to fertilize crop fields. A third type developed in India uses five or six fish species in the same pond. Fish species are selected based on different food habits so they do not compete using surface, column, and bottom feeders. The system is very productive. A fourth type involves use of large plastic tanks in a greenhouse. Tilapia are grown in the tanks and feed on algae. The tilapia waste is collected in the hydroponic bed and used to fertilize commercial crops.

A fifth system is classic fry farming where trout or other sport fish are raised from eggs to fry or fingerlings and transported to streams or ponds for release. They are raised in long shallow concrete tanks fed with fresh stream water. The fish are fed commercial fish pellets.

Good management of commercial fish-raising operations requires clean water, suitable facilities, knowledge of fertilizer application, feed, aquatic weed control (in some cases), and prevention of undesirable fish invasion as well as parasites and diseases.

Chapter 11. Literature Cited

Anonymous (2016). Fish Farming. From Wikipedia, the free encyclopedia. Retrieved from https://en.wikipedia.org/wiki/Fish_farming. Accessed 17 August 2016.

Waters R, Kelly HD, Dollar WM (1978). You Can Grow Fish For Fun or For Profit. Pages 385-395 in Hayes J ed. Living on a Few Acres. The Yearbook of Agriculture 1978. Superintendent of Documents, U.S. Government Printing Office. Washington DC.

Meyers FP (1991). Aquaculture Disease and Health Management. Journal of Annual Science 69:4201-4208.

FOOD SCIENCE

Introduction

So far in this book we have discussed the tools, crops, animals, and resources needed to provide human food, animal feed, energy, shelter, clothing, and aesthetics. We now need to discuss food preparation, food safety, and preservation.

Food science is the study of the physical, biological, and chemical makeup of food and the concepts of food processing. Food technology is the application of food science to the selection, preservation, processing, packaging, distribution, and safe use of food.

McWilliams (2001) indicated that the modern food market is a complex and ever-changing situation and that satisfying consumers is an ever-changing and daunting task. The shift in population age, diversity, lifestyles, and health concerns impact the types of food consumed. Cancer and heart disease may prompt consumers to eat more fruits, vegetables, and cereal products. Some consumers demand organically-grown food while others are skeptical of genetically-modified organisms (GMO) even though genetic engineering of the animal or plant species may improve the nutrition and food quality. Alternately, GMO foods may cause an allergic reaction although plant and animal hybridization has been practiced for many years. In the

U.S. the U.S. Food and Drug Administration (FDA) presently does not require labeling of GMO products.

Examples of improved food products as a result of genetic engineering are both exciting and beneficial. Commercial viable products include increased essential amino acid content in corn and soybeans, potatoes, and tomatoes with higher solids content and uniform ripening of fruit and vegetables for market, just to name a few.

Experimentation with food originated from ancient customs and ceremonies with eventual observations and research in the home. Nowadays technological advances or changes are a result of food preparation in processing plants and factories as well as the home.

Why Do Research in Foods?

First, basic nutrition of foods is paramount to good human health. Each food type presents opportunities and problems. Plants and animals are researched to improve yield, vigor, nutritive value, and environmental quality while attempting to maintain efficiency of agriculture production and costs. All phases of scientific research are conducted and the results are published in scientific journals, popular magazines, and on the internet. Trained scientists and skilled experts work to improve and investigate all types of crop plants and animals to improve and sustain agriculture, including food science. This data is disseminated to the public by radio, television, online, and through the printed word, food specialists, food packaging, and many other forums.

Types of Topics Covered
A. Milk and dairy products
B. Nutrient content of vegetables
C. Food microbiology
D. Nutrient value of poultry and meats
E. Effects of freezing
F. Carbohydrate content of cereals
G. Baking properties of potatoes
H. Chemical contaminants
I. Natural toxicants in foods
J. Basic food chemistry

Food Evaluation

Food quality is evaluated by all people whether consciously or unconsciously. Food is purchased in the marketplace by consumers based on previous experiences or by recognized brands. Food testing is done by sensory or subjective evaluation or objective evaluation (McWiliams 2001).

Sensory Evaluation

McWilliams (2001) indicated food products can be tested in-house by a company or manufacturer, but further testing should be done by a focus group of 4 to 12 typical consumers who fit the specific demographic characteristics of interest to the food company. Large-scale (200 to 500 consumers) testing can be done at a central location or in-home testing. These tests provide valuable sensory information for consumer acceptance. These tests are often market tested on a small scale to determine their potential and the sensory and objective evaluation may determine if modifications are needed or if the product should be discontinued in the marketplace.

Sensory evaluation consists of odor perceptions, taste, and visual messages. Sensory evaluation includes all the senses as they come in contact with food. Visual evaluation includes judgements of color as well as contour texture. The olfactory sense evaluates the aroma and perception of flavor. The taste buds in humans can identify sour, sweet, salt, and bitter components of flavor. Tactile (touch) is important as is crispness. Even auditory cues may be a part of food evaluation.

Scorecards for food evaluation panels can be developed for the specific experiments targeted and the information needed. Food samples then can be ranked to help the food industry improve, accept, or reject specific food items.

Objective Evaluation

Objective evaluations of food can be extremely important and complement the sensory or subjective evaluation. Many physical measurements with a wide range of devices can be used to measure such factors as volume, specific gravity, moisture or juiciness, texture, tenderness, color, and flowability in foods.

Chemical measurements also include nutrient analysis and pH. Sophisticated flavor research utilizes gas-liquid chromatography (GLC) or high-pressure liquid chromatography (HPLC) to separate volatile flavor chemicals from the total flavor sample. Purified compounds from the mixture are then identified using infrared spectrophotometry and sometimes mass spectrometry. The goal is to identify the key natural flavor components and formulate synthetic flavors that match natural flavors in foods.

Carbohydrate, lipid (fat), and protein content of foods can be done chemically following professionally recognized and approved methods (McWilliams 2001).

Physical and Chemical Characteristics of Food

Water

Water is a common component of foods and food preparation. Water can be found in solid, gaseous, or liquid form in foods. One of the reasons water content is so important in foods is that it influences food spoilage and safety. Certain microorganisms toxic to humans may flourish, if allowed, in many fresh foods including meat, milk, and vegetables at certain moisture levels. Dried foods at moisture levels of 12 percent or less are not susceptible to microbiological spoilage. Water activity in food can be influenced by drying, freezing, and adding solutes such as salt or sugar. Water content of foods can significantly affect its presentation and acceptance. Cooking with soft or hard water (salts of calcium and magnesium carbonate or sulfate) influence food products. For example, hard water can cause cloudiness in tea and coffee.

For health reasons safe water free of harmful chemicals, microorganisms, or any component detrimental to health should be avoided. One part per million of added chlorine is recommended.

Carbohydrates

Carbohydrates are hydrates of carbon. They can be simple sugars or complex carbohydrates. They are called mono- and disaccharides (simple sugars) and polysaccharides (complex carbohydrates).

The sweet taste of these compounds in foods contribute to acceptability. Carbohydrates are composed of carbon, hydrogen, and oxygen. The hydrogen and oxygen are usually present in about 2:1 ratios as found in water. Familiar pentoses include ribose, arabinose, and xylose. Common hexoses are glucose, fructose, and galactose. Common disaccharides (two saccharide units joined together) include sucrose (glucose and fructose), maltose (two units of glucose), and lactose (glucose and galactose).

Sugars

Oligosaccharides contain between three and ten glucose units but are not common in foods but form as polysaccharides that are hydrolyzed into basic components. Polysaccharides are very large polymers of saccharides joined by 1,4 linkages and occasionally by 1,6 linkages. Dextrins, cellulose, and two fractions of starch (amylose and amylopectin) are common polysaccharides in foods. Pectin and gum are polysaccharides usually as galactose derivatives; gums sometimes contain other sugar derivatives as well.

Most of the sugar products on the market are derived from sugarcane or sugar beets and refined to the desired purity.

Starch

Starch is composed of a linear fraction, amylose and a highly-branched fraction of amylopectin. They are deposited in granules in the leucoplasts of some cereal grains and roots of some plant food sources. About 75 to 80 percent of the granule is amylopectin and the remainder is amylose (McWilliams 2001).

Fiber

Fruits and vegetables provide a variety of flavors, nutrition, and fiber. Structural components include cellulose, hemicellulose, pectic substances (protopectin) (pectin, pectinic acid and pectic acid) and lignin. Parenchyma cells predominate in the fleshy portions of fruits and vegetables. The gums found in fruits and vegetables are complex carbohydrates that are used to modify foods. They act as thickening agents, fat replacements, and texturizing agents.

The pigments in fruits and vegetables include chlorophyll, the carotenoids and flavonoids. Color changes occur because of heat, change in pH, oxidation, and enzyme action (McWilliams 2001).

Fats and Oils

Lipids, like carbohydrates, are organic compounds composed of carbon, hydrogen, and oxygen. Oxygen, however, occurs in much smaller proportions and hydrogen in larger proportions in lipids than in carbohydrates. Lipids have more energy value than carbohydrates.

In food, key components are glycerol and fatty acids. Combined they form an ester. When glycerol is esterified (formation of an ester from an alcohol and an acid), the polyhydric nature of glycerol permits the formation of a wide range of simple fat molecules. Each of the hydroxyl groups can esterify with a different fatty acid making possible molecules of simple fats extremely large. Food fats contain a relatively large and varying content of different fatty acids. However, the glycerol molecule remains a constant structural feature.

Functional roles of fat in food are color, flavor, texture, tenderness, emulsification, and a cooking medium. Fat replacements are being widely used in commercial food to reduce calories.These replacements may be protein-based, carbohydrate-based, or fat-based (McWilliams 2001).

Proteins

Proteins are composed of amino acids that have an amino group and an organic acid radical. Amino acids join through peptide linkages between the nitrogen of one amino acid and the carbon of the carboxyl group in the next amino acid to form the primary structure. These various organic compounds (designated as R groups) range from the simple hydrogen atom of glycine to dual ring structures such as in tryptophan. There are 22 amino acids with nine essential amino acids for life (McWilliams 2001).

The primary structure is coiled into a helical configuration. The secondary structure is in a relaxed shape or other forms because it is held by secondary bonding forces. Superimposed on the secondary structure is a distorted tertiary structure held by secondary bonding forces. Very rarely are proteins in foods found in a quaternary structure. Some proteins in food are fibrous or conjugated but usually globular in shape. All enzymes, some hormones, and oxygen transporting proteins are globular.

When heat or other energy is applied to food containing protein, the protein may denature and relax from the tertiary and secondary, low energy structure. With continual energy input the molecules may result in precipitation and thicken with loss of solubility. Coagulation may also occur causing clumping. This can be used in food preparation like forming egg white foams and gelatin foams. Egg white foams can be used to give stable food products and thickening agents. Milk proteins can be precipitated to form curd used in cheese making. Gluten provides the structure for baked products.

Food Groups

The five food groups according to the Australian Government, Department of Health are:

1. Vegetables and legumes/beans
2. Fruit
3. Grain (cereal) food, mostly whole grain and/or high cereal fiber varieties

4. Lean meats and poultry, fish, eggs, tofu, nuts and seeds and legumes/beans
5. Milk, yogurt, cheese and/or alternatives, mostly reduced fat (Anonymous
2015).

A balanced diet includes a variety of foods from each of the five food groups usually daily. Fruit and vegetables provide vitamins, minerals, and fiber. Milk, yogurt or cheese provide calcium for healthy bones and teeth. Lean meat, fish, poultry, eggs, nuts, and legumes are essential for growth of children. Meats are rich in protein, iron, and zinc. Children's diets should be limited in fats.

Vegetarians should eat a variety of legumes, nuts, seeds, and grain-based foods to gain the same nutrients that meat, poultry, and fish provide (Anonymous 2015).

Wikipedia, the free encyclopedia (Anonymous 2016) food group is similar to the Australian Government Department of Health list as follows: dairy (milk products), fruits, grains, beans and legumes (cereals), meat, confection (sugary foods), vegetables and water. Alcohol drinks are listed separate from other food groups and recommended only by and for certain people in moderation.

Food Preparation

The nutritional quality and quantity of plant and animal products were discussed in earlier chapters of this book. However, food preparation and cooking can change the chemistry, structure, and nutritional qualities of the prepared food. For example, fluid milks are not able to form stable foams; but undiluted, chilled evaporated milk and nonfat dried milk solid diluted with an equal volume of water can be whipped into foams. These foams need to be stabilized with gelatin or gums if they are to be used in food products (McWilliams 2001).

Gelatin is the protein derived by extraction of collagen from animal skins and bones. A gel forms as the gelatin molecules establish a continuous, solid network by forming hydrogen bonds and other secondary bonds. Water is bound to these molecules and trapped within the framework. Gelatin gels and sols are reversible, depending on the temperature. When just beginning to congeal, it can be beaten into a useful, light foam for use in desserts and salads (Mc Williams 2001).

Eggs are emulsifiers, thickening agents, and foaming agents. Products that illustrate these different roles include hollandaise sauce, mayonnaise, cake batters, cream puffs, custards, cooked salad dressings and sauces, pie fillings and cream puddings, and so on (McWilliams 2001).

Cereal grains can be milled into flour. In baked products, flour is used to provide the basic structure. Liquid hydrates the gluten, dissolves the sugar and other dry ingredients, gelatinizes starch, and provides steam for leavening. Eggs provide liquid, protein, emulsification, flavor, color, and a lighter texture when added as a foam. Fats, sugar, and added salt complete the baked product (McWilliams 2001).

Commercial Food

Jack Allen's Kitchen celebrates the tastes of Texas in Austin (Gilmore and Dupuy 2014). Jack Allen's Kitchen is an example of a restaurateur with extensive culinary experience working with local growers and producing a long list of foods and recipes. The restaurant uses local products in about every dish served, but products like salmon come from other sources. The culinary list includes cocktails, appetizers, side dishes, dressings, entrees, and desserts. The list is too long to print here but the restaurant feeds hundreds of people daily, probably including some of his own suppliers. There are hundreds more restaurateurs in the U.S.A. that provide both fast and regular food products for those wishing to eat away from home.

Homemade Food

In contrast, Belanger (2009) provides instructions on growing and producing food in a self-sufficient style for self-sufficient living. Belanger (2009) writes how to grow vegetables, fruits, and livestock to save money, how to improve nutrition by avoiding fast food, and how to become self-sufficient. Most people cannot or do not want to entirely grow their own food but could benefit from less processed food. Belanger (2009) recommends a well-stocked pantry and storing food in root cellars, home canning dry foods (dehydration) or freezing them. Belanger (2009) provides recipes for homemade peanut butter, roasted peanuts, and sesame butter. He also provides pasta, bread, yogurt, and cheese recipes made at home.

Specialized Food

There has been an explosion of freeze-dried food makers in recent years with dozens of companies offering hundreds of products (Draper 2016). These freeze-dried meals offer the survivalists, soldiers, campers, hunters, fishermen, and others easy access and a variety of nutritious foods. They offer simple, just add water preparation of high-calorie light-weight packages. Some companies offer both freeze-dried and dehydrated foods. Freeze-dried food is made by placing it under vacuum and super-cold temperatures where the water vaporizes and dissipates. Dehydrating food is done by exposing food to low to moderate heat and drying it.nDrying food can be done in a standard home oven. The grocery store and specialty stores sell dehydrated foods such as fruits, vegetables, and other ingredients. Freeze-dried and dehydrated food is a good way to preserve foods with fast access in lightweight, handy packaging.

Natural Toxicants in Foods

Watson (1998) describes natural occurring chemicals toxic to humans arising from production by living organisms. Natural toxicants in food can originate in plants, bacteria, algae, fungi, and animals. They can be in unprocessed or processed foods. Farm animals can, in some cases, act as a biological barrier between natural toxicants in animal feed and meat-eating or milk-drinking consumers. This depends on whether or not the animal can detoxify or excrete the toxicant from the food likely to be consumed.

The routes by which we are exposed to bacterial toxins is more direct and affects our gastrointestinal tracts. Toxicants from algae can contaminate our food by one main route, directly up the food chain; for example, from toxigenic (toxin-producing) algae that are consumed by some filter-feeding mollusks which we, in turn, consume. However, natural toxicants from fungi are probably the most ubiquitous. Mycotoxins produced by some fungi can contaminate our food at about any stage of production.

Possible Effects of Natural Toxicants

Bacterial and algae toxins usually cause relatively immediate and short-term illness in the form of indigestion, stomach cramps, diarrhea, etc., whereas some natural toxins may cause more long-term illnesses such as cancer. However, with these toxicants it can be sometimes difficult to establish the link between cause and effect. Research continues on this important subject.

Examples of Natural Toxicants

Psoralens are found in many plant families including Umbelliferae, Leguminosae, and Rutaceae (citrus fruits) and have been implicated in adverse reactions to some antihistamine preparations. Levels are reduced by cooking, especially boiling.

Cyanogenic and other glycosides found in stone fruit kernels and lima beans are not toxic themselves; but on hydrolysis, hydrocyanic acid (HCN) is released. Human toxicity from HCN is rare but livestock losses have occurred when forage sorghums and wild cherries are eaten.

Bracken fern is consumed in Japan. Boiling in water removes the toxin in young ferns. Bracken fern has caused cattle losses when consumed.

Glycoalkaloids toxicants are found in greened potatoes and at lower levels in normal potatoes. High levels of these alkaloids cause gastroenteritis. A tolerance of 20 mg of alkaloids per 100g of fresh tuber weight has been generally accepted for consumption. New varieties to reduce these alkaloids are being investigated.

Other Contaminants in Food

Watson (1993) lists chemical contaminants in food that can originate from crop and animal production, food manufacturing, packaging, storage, and industrial procedures. The contaminants include pesticides, nitrite, nitrate, metals, naturally occurring toxicants, veterinary drugs, and environmental organic chemical contaminants.

Microbiological hazards can also occur from bacteria, viruses, protozoa and parasites and bovine spongiform encephalopathy (BSE) (Roberts 2001). However, every effort is made in our food supply to reduce or eliminate chemical and natural occurring contaminants in food through careful growing of plants and animals, processing, food preparation, manufacturing, storage, and packaging.

Summary

Food science is the study of the physical, biological, and chemical makeup of food and the concepts of food processing. Food technology is the application of food science to the selection, preservation, processing, packaging, distribution, and safe use of food.

The modern food market is a complex and ever-changing situation to satisfy consumers. Markets attempt to produce appealing products for the home and commercial preparations with satisfactory nutrition. The basic food groups include dairy products, fruits, grains, beans and legumes (cereals), meat, confection (sugary foods), vegetables, and water.

Food is evaluated by sensory and objective means. Sensory tests of food consist of odor perception, taste, and visual messages. Sensory evaluation includes all the senses. Visual evaluation includes judgements of color, contour, and texture. The olfactory sense evaluates the aroma and perception of flavor. The taste buds in humans identify sour, sweet, salt, and bitter components of flavor. Touch is important as is crispness. Even auditory cues may be a part of food evaluation.

Objective evaluation of food includes many physical measurements such as volume, specific gravity, moisture or juiciness, texture, tenderness, color, and flowability. Chemical measurements include nutrient analysis and pH. Sophisticated instrumentation can measure volatile flavor. Carbohydrate, lipid (fat), and protein content of foods can also be done by recognized chemical procedures.

Water content, type of carbohydrate, fiber, fats and oils and proteins are important components of food. Preparation of homemade, commercial, and specialized food is discussed, as well as natural and man-made contaminants of food. However, every effort is made to ensure a safe and abundant food supply in the U.S.A.

Chapter 12. Literature Cited

McWilliams M (2001). *Foods: Experimental Perspectives.* 4th Edition. Prentice-Hall, Inc., Upper Saddle River, New Jersey. 503 p.

Gilmore J and Dupuy J (2014). *Jack Allen's Kitchen: Celebrating the Tastes of Texas.* University of Texas Press, Austin, Texas. 327 p.

Anonymous (2015). The Five Food Groups. Australian Government. Department of Health. National Health and Medical Research Council. Retrieved from https://www.eatforhealth.gov.au/food-essentials/five-food-groups. Accessed 10 October 2016.

Anonymous (2016). Food Group. Wikipedia, the Free Encyclopedia. Retrieved from https://en.wikipedia.org/wiki/Food_group. Accessed 10 October 2016.

Belanger JD (2009). *The Complete Idiot's Guide to Self-Sufficient Living.* Penguin Group (U.S.A.) Inc. 375 Hudson Street, New York, NY. 379 p.

Draper D (2016). Backcountry Grub. Petersen's Hunting. October 2016, Volume 44, Number 3. Outdoor Sportsman Group, New York, NY. Pages 40, 42, 44 and 46.

Watson D (1998). What are Natural Toxicants? Pages 1-8 in Watson D ed. Natural Toxicants in Food. Sheffield Academic Press, Sheffield, England.

Watson D (1993). Introduction pages 1-7 in Safety of Chemicals in Food: Chemical Contaminants. Watson D ed Ellis Horwood Limited. Chichester, West Sussex. PO 19 1EB, England.

Roberts CA (2001). *The Food Safety Information Handbook*. Oryx
Press. Westport, CT. 312 p.

PEST CONTROL

Introduction

Since earliest times humans have had to compete with wildlife, weeds, insects, nematodes, plant and animal pathogens, rodents, and other pests for their food supply.

Early methods to control pests were limited to hand picking, fire, use of salt or ashes, or just plain physical cutting or physically subduing plant or animal pests. These methods were very temporary and usually didn't solve the pest problem.

As humans developed during the centuries, humans recognized some plants and animals were more resistant to certain weeds, insects and diseases; and they were selected and propagated. Timing of planting, select location, select species, and appropriate harvest helped produce better crops and animals. Improved sanitary conditions and protection from adverse weather and selection of superior animals improved animal production. Fire was used to control weeds, improve pastures, kill insects and diseases, and open land for crop production.

Today we have excellent cultural treatments for crop production and chemicals that control weeds, insects, plant diseases, and nematodes. We also have improved genetics in plants that provide vigor, good nutrition, insect and disease resistance, drought resistance, and many other desirable features. In animal production we have more productive animals and insect- and disease-resistant species that provide abundant meat, milk, and eggs of superior quality.

The genetics are also being improved in animals to improve vigor, heat and cold resistance, nutritional qualities, and other characteristics that are needed for success in animal production.

Pesticide Industry Sales and Usage

World pesticide expenditures totaled more than $35.8 billion in 2006 and more than $39.4 billion in 2007 (Grube et al. 2011). Expenditures for herbicides (40%) accounted for the largest portion followed by insecticides, fungicides and other pesticides, respectively. U.S. expenditures totaled $11.8 billion in 2006 and $12.5 billion in 2007 in proportions similar to world expenditures with a larger proportion spent on herbicides. The U.S. in 2006 spent 5.7, 4.1, 1.2, and 0.9 billion dollars on herbicides, insecticides, fungicides, and other pesticides, respectively. In 2007, it was 5.8, 4.3, 1.4, and 0.9 billion dollars on herbicides, insecticides. fungicides, and other pesticides, respectively.

In 2007 the percentage usage of herbicides was 72, 15, and 13, respectively, for agriculture, industry, and commerce/government and home and garden. For insecticides it was 46, 17, and 38% for agriculture, industry, and commerce/government and home and garden in 2007, respectively. For fungicides it was 78, 18, and 5% for agriculture, industry, and commerce/government and home and garden, respectively. Other pesticides in 2007 usage were 67, 8, and 25% for agriculture, industry, and commerce/government and home and garden, respectively.

Pest Control Methods

These data show the large usage of chemicals for weed, insects, plant pathogens, and other pests in the world and U.S.; but there are also other methods that are extremely useful. Other methods include mechanical, biological, and sometimes prescribed burning. Prevention of weed, insect, and plant pathogen problems before they become a big problem is many times possible and the best approach.

Minor use of flame cultivation to control weeds has been used in cropland mainly with butane or propane when weeds are small in cotton, but is rarely used today. The effective herbicides available and high cost of natural gases are primarily responsible for the loss of interest in flame cultivation (Buchanan 1992).

Cultivation or tillage to control weeds is still very important and essential for planting and weed control maintenance in row crops. Biological control of weeds and insects is practiced on a limited scale in U.S. agriculture.

Biological Control

Biological control is a method of controlling pests such as insects, mites, weeds, and plant diseases using other organisms. It relies on predation, parasitism, herbivory, other natural mechanisms and occurs constantly in nature that limit the growth of about every living organism or the potential to limit about every living organism (Anonymous 2016a).

If one of these biological control methods is useful in animal or plant production, it may be useful in controlling pests of that crop or animal. It can be a part of integrated pest management (IPM) if used with other control methods.

Three basic types are (1) releasing a natural enemy, (2) augmentation (breeding and releasing natural enemies) to improve control and (3) conservation by using other methods to increase natural enemies of the pest. Natural enemies of insect pests are predators, parasitoids, and pathogens. Biological control agents of plant diseases are referred to as antagonists and can be bacteria, fungi, viruses, and algae. Biological control agents of weeds include seed predators, herbivores, and plant pathogens.

Chemical Control

Chemicals such as sulfur and salt were used in antiquity for insect and weed control respectively. The Romans are known to have used salt to destroy the crops of their enemies and Democritus, the Greek philosopher, advocated applying olive oil to reduce certain diseases of plants (Ware and Whitacre 2004).

The Chinese used arsenic as an insecticide by 900 AD. Pyrethrum, lime and sulfur combination and arsenic, sulfur and soaps were found effective for pest control by the early 1800s. By 1825-1850 quassia, phosphorous paste, and rotenone were employed. Mercurous chloride was used as a seed treatment during this time. By 1865 control of downy mildew on grape vines by copper sulfate, lime and water (the Bordeaux mixture) was successful in France and is still used today (Ware and Whitacre 2004).

These advances in chemical pest control of the 19th century have led to a tremendous list of useful pesticides for use on our crops, homes, and animals. Worldwide losses from insects, diseases, weeds, and rodents were estimated at more than $100,000 billion annually in 2004. Pesticides have become indispensable tools for mankind (Ware and Whitacre 2004).

Modern Beginning

The beginning of the modern era of pesticide development using organic chemicals got off to a good start with the development of DDT (insecticide) before World War II in 1939, the synthesis of 2,4-D (herbicide) in 1941 by Pokorny and the emergence of the first patent of dithiocarbamate fungicide (zineb) in 1943.

Since the early 1940s we have had an explosion of new products from a variety of manufacturers and companies. Many new materials have replaced old ones for many reasons. Newer, more environmentally friendly materials have replaced older ones. Some pesticides have become problems because of cost, toxicity, too persistent or loss of effectiveness. Some weeds or pests have become resistant to commonly used materials, and control strategies must be changed.

Why Pesticides?

Plants on which humans and other animals depend for life are susceptible to about 100,000 diseases caused by viruses, microorganisms, and other plants. Some 30,000 weed species the world over compete with field crops. Many weed species (1800) cause serious economic loss and must be controlled. Among the one million insects, about 10,000 contribute to devastating losses in crops worldwide.

More than 1000 to 3000 nematode species also cause severe economic loss to crops. Many of these pests cannot be controlled or managed by other means so pesticides are a must (Ware and Whitacre 2004).

In addition to weeds, insects, and plant and animal diseases, certain problem snails and slugs, fish, birds, wildlife, rodents, varmints and other pests need to be dealt with by some control method or repellants.

Continual Awareness

If possible it is desirable to recognize a weed, insect, or pest problem early or early in the life cycle of the pest and treat or control it before it becomes a big, costly problem. The second point is that a one-time chemical or mechanical treatment may not be possible since that pest problem may reoccur annually or new problems may emerge of a different pest species or a resistant pest of the same species.

In the following pages examples will be given on management of weeds in crops and rangeland, insect control, management of animal and plant diseases, and management of other pests of urban sites and agriculture.

Examples of Pest Management

Corn - Weeds

The United States is a major producer of corn (see Chapter 4). Herbicides can be an effective weed control measure in corn, but the type of weeds present dictate the type of herbicide used. Growers should not depend on herbicides alone, but integrate good cultural practices so the corn plant will be competitive with all weeds. Growers can integrate chemical weed control with cultivation, especially on difficult-to-control weeds or when weather conditions reduce herbicide effectiveness.

The first step in good cultural practices is selection of a good corn hybrid adapted in the area in which it is grown (Jemison and Bhowmik 2007). Other considerations are potential weed problems, crop and herbicide rotation, tillage systems, soil properties and potential environmental risk and cost. Herbicides need to be rotated (changed) to avoid development of herbicide resistance in weed populations. Several herbicides are available for use. Genetically-modified (GM) corn varieties are available resistant to herbicides such as Roundup (glyphosate) for use. Herbicides can be applied before the weeds emerge (pre-emergence) or after the weeds emerge (post-emergence) depending on the type of weeds present. Some herbicides can be used either pre-emergence or post-emergence for weed control, but many times they are specific for one use only. Specific herbicides vary in persistence in the environment so care must be taken in potential risk to the next crop in the rotation. For example, corn (grass) may tolerate the herbicide used but many injure soybeans (broadleaf plant) in the next year's rotation if improperly selected.

The herbicide label on the manufacturer's container will list the weeds controlled, the amount of herbicide to apply, and the stage of corn growth at time of application. Herbicide approval for each crop is under EPA regulations.

Corn - Insects

Pillbugs, white grubs, wireworms, seed corn maggots, cutworms, sugarcane beetles, stink bugs, chinch bugs, southern corn rootworms, and western corn rootworms are major seed and seedling pests of corn. Insecticides are available to control them in Tennessee and other locations. They are treated with insecticides when these insects reach critical numbers on the corn plant (Anonymous 2016b).

When corn is in the whorl-stage or larger, fall army worms and corn earworms, European corn borer, southwestern corn borer, Japanese beetles, and stink bugs are also treated when the numbers justify insecticide applications.

The corn plants should be scouted continually for insect pests and damage to the corn plants.

Corn - Diseases

The American Phytopathological Society (Shurtleff et al. 2016) lists 12 bacterial diseases, 15 fungal diseases plus nearly 60 downy mildews, a mite, 16 parasitic nematodes, 2 parasitic higher plants (witchweed) and numerous virus and virus-like diseases. Within each of these genera mentioned, one or more species may occur on corn.

Corn seedling diseases are seed rot, seedling blight, and root rot. The most common diseases are caused by *Phthium, Fusarium, Gibberella, Trichoderma* and *Panicillium* but fungi such as *Diplodia* and *Rhizoctonia* can be involved. They all live and thrive in the soil and can affect other crops besides corn (Anonymous 2016c).

Corn leaf diseases involve *Anthracnose* leaf blight and can be a problem in warm, wet years. Symptoms are leaf spotting, top dieback, and stalk rot. Control is helped by resistant hybrids, but cultural treatments using crop rotation (avoiding second-year corn) is best. Fungicide treatment may not be economical unless it is a high value seed corn crop.

Other corn leaf diseases include northern leaf blight, eyespot, bacterial leaf blight, gray leaf spot, common rust, common smut, and head smut.

Northern leaf blight has been one of the most damaging corn leaf diseases. Use of hybrids resistant to the common races of northern leaf blight should be used, plus crop rotation.

Eyespot usually causes minor losses in corn. Resistant varieties, crop rotation, and clean plowing of crop debris help control the disease.

Bacterial leaf blight is a concern in southwestern Ontario and Essex and Kent Counties where the majority of seed corn is produced. The disease is controlled by managing the corn flea beetle which carries the bacteria.

Grey leaf spot can be destructive. Crop rotation and a resistant corn hybrid may be necessary.

Common rust does not overwinter in Ontario, Canada. It originates from infected corn in the southern U.S. and Mexico. Common corn hybrids should be grown. Common smut and head smut occur in Ontario, Canada and can be severe causing 25% of the plants to have smut galls. Most commercial com hybrids have resistance to smut; however, it can still be a problem in corn fields.

Stalk rot includes *Anthracnose* stalk rot, *Gibberella* stalk rot, *Fusarium* stalk rot, *Diplodia* stalk rot, and *Pythium* stalk rot. Commercial corn hybrids, seed corn inbreeds, and specialty corn hybrids are used. Managing insects like the European corn borer, good weed control, proper corn population, soil fertility, crop rotation, and proper tillage help control the corn stalk rots.

Ear rots and moulds occur. Harvest as early as possible, cool grain after drying, clean storage bins, and check for temperature and wet spots, insects and mould, exercise feeding affected corn to livestock and test for toxins with mould-affected grains (Anonymous 2016c).

This long treatise on corn plant diseases demonstrates the seriousness of planting resistant corn hybrids and cultural practices to limit these many disease problems. Agricultural researchers work long hours to develop cultural and chemical treatments and corn hybrids to ensure our continued food and feed supply. The same can be stated for management of weeds and insects in corn and all our crop plants. What is stated for Canada can also apply to the United States in corn production.

All crops grown in the United States and worldwide may have weed, insect, and plant disease problems including soybeans, wheat, barley, grain sorghum, cotton, rice, and many others. The crop manager or farmer must be aware of any weed, insect, or disease problems and scout the fields often to manage these pests before they devastate the crop even if organically grown. Methods are available to handle just about any pest problem in horticultural or agronomic crops.

Alfalfa - Weeds

Alfalfa is used as an example of a pasture forage that can be grazed or used for hay. Weeds are a serious problem in alfalfa seedling establishment. In the central United States (Kansas) several options of herbicides exist for alfalfa establishment, including pre-plant incorporated and post-emergence herbicides. Once established, several post-emergence herbicides can be used depending upon the weed encountered. Fall seeding or spring seeding can be done. In Kansas weed problems in fall seeding include cheatgrass, mustards, volunteer wheat, henbit, and field pennycress. Spring seeding may have pigweed, foxtail, velvetleaf, kochia, and crabgrass problems. In alfalfa seedling establishment, good seedbed preparation is needed to eliminate weed problems where possible before any herbicide use (Anonymous 2016d).

In established alfalfa, summer and winter annual weeds just described for alfalfa establishment can be problems; but greater problems may be perennial weeds such as johnsongrass, bindweed, and curly dock. Herbicides can be used to control these weeds. Also during the dormant season, light tillage has been used in central and western Kansas to control winter annual weeds without any apparent long-term damage to the alfalfa.

Alfalfa can be irrigated where applicable 5 to 6 days before and after mowing.

Herbicides used can be recommended by the local county extension agent, state university extension service, industry, agricultural consultation, or the U.S. Department of Agriculture.

Alfalfa - Insects

Kansas State lists early spring pests in alfalfa as army cutworm, blue alfalfa aphid, cowpea aphid, pea aphid, alfalfa weevil, and the clover leaf weevil. Summer and fall pests include several worms: the alfalfa caterpillar, beet armyworm, fall armyworm, variegated cutworm and webworms. The cowpea aphid and spotted alfalfa aphid can also occur. Miscellaneous pests include blister beetles, grasshoppers, potato leafhopper, and seed chaliced (Anonymous 2016e).

Alfalfa - Diseases

Hanson and Barnes (1975) indicated bacterial wilt caused by *Corynebacterium insidiosum* is one of the most destructive alfalfa diseases in the U.S. and occurs in practically every state. Affected plants are stunted and eventually die. The disease is controlled by resistant cultivars.

Anthracnose (*Colletotrichum trifolii)* is the cause of summer decline in the eastern U.S. Large sunken lesions appear on stems near the soil surface, dead shoots, and crown root. Resistant cultivars are used.

Leaf diseases caused by *Pseudopeziza medicaginis, P. jonesii, Leptosphaerulina briosiana* and *Stemphylium spp.* and other organisms such as rust (*Uromyces striatus*) and downy mildew (*Peronospora trifoliorum)* lower hay quality.

Spring black stem and leaf spot (*Phoma medicaginis* and *Ascochyta imperfecta*), crown and stem rot (*Sclerotinia trifoliorum*), fusarium wilt and root rot (*Fusarium spp.*), Phytophthora root rot (*Phytophthora megasperma*), dwarf disease (a virus), and crown wart are other diseases that may cause serious damage (Hanson and Barnes 1975).

Alfalfa serves here as an example of the weed, insect, and disease problems that can occur in a crop that is grazed by livestock and wildlife or pelleted or made into hay for supplemental or winter feed. Alfalfa is a very valuable crop in the United States for livestock and poultry.

Pasture and Rangelands - Pest Management

Common insects on pastures and rangelands include army cutworm, army worm, blackgrass bugs, cereal leaf beetle, grasshoppers, and Mormon crickets in the western United States. Some of these same insects feed on crops. When warranted there are a number of insecticides labeled for these insects.

In Texas grasshoppers, armyworms, and fall armyworms are the common insect pests of pastures. In certain sites of the state the desert termite has occasionally damaged grasses. East Texas fire ants may often give hay producers trouble in harvest, shipping, and loading hay crops. The ants build mounds that can break machinery when cutting hay. Disk-type cutters instead of sick-bar-type cutters in ant-infested areas can be used.

In North Carolina chemicals are used to control fire ants, but there is no single simple solution for managing fire ants on a farm. There are only a few chemicals labeled for pasture use.

The biggest problem of pastures and rangelands is herbaceous and woody weeds that rob the land of productive forage for livestock. Several herbicides are available and labeled for both weeds and undesirable brush. Shredding or mowing is also done to remove weeds and small brush temporarily from grasslands. Prescribed burning is also done where permitted if the proper equipment and personnel to legally prescribe burn are present. Mechanical methods, in addition to shredding, are chaining, root plowing, grubbing, bulldozing, and can be employed in larger brush and weeds. Aerial spraying of herbicides in tall vegetation and rough terrain is sometimes employed (Bovey 1987, 2016).

Combinations of herbicide application or mechanical treatments followed by prescribed burning several years after the initial application help maintain forage species and weed and brush control. Other treatment combinations are possible. Biological control of weeds and brush on rangelands has had limited use whether using insects, plant pathogens, selective grazing, or plant competition.

The holistic approach more recently is to leave enough weed and brush on rangelands to promote both wildlife and domestic livestock.

Forest Land - Pest Management

Forests, like cropland, need to be protected from weeds, insects, and disease problems. The threat of insects destroying vast areas of forests are real according to a recent Associated Press article (Casey and Whittle 2016). They state 63 percent of U.S. forests are at risk through 2027 and will cost several billion dollars annually in dead tree removal, reduced property values, and increased timber losses. According to forest scientists, hundreds of pests have invaded the nation's forests and the emerald ash borer alone has the potential to cause over $12 billion in damage by 2020.

The primary reason for the invasive pest problem is globalization from increased travel and trade. The emerald ash borer found in 2002 in Michigan is now in 30 states and has killed millions of ash trees. The gypsy moth discovered in 1869 in Boston is found in 20 states and has reached as far north as the Great Lakes. Native bark beetles have spread from Mexico to Canada.

To control the ash borer, certain wasp species have been released to combat them but it may not be effective enough to control them.

Researchers are also considering genetic modifications of the trees to make them more resistant to insect attack.

Some insect pests are bark borers, defoliators, wood borers, tip feeders (attach young twigs), sap suckers, and root feeders. These insect pests weaken the tree and secondary invasion can be by a plant disease such as fungi, bacteria, nematodes, viruses, and mistletoes; but of the parasites listed, fungi are most destructive.

Chemicals are available to kill insects and plant diseases but must be approved by EPA. Chemicals may be effective on small areas when the problem is first recognized, but treatment of large areas of forests may be too costly. The insect or disease problem must be identified by an expert to make sure of the correct problem so the proper treatment can be applied whether biological or chemical.

Weeds and brush are also a big problem in forest management. There are many weeds that can be problems and must be controlled if the forest is to produce to its full potential. Weed control can be done with chemicals, mechanically, or by prescribed fire or certain combinations of these methods. Control of small weeds and brush can be done chemically or mechanically between the desirable trees followed by prescribed burning for maintenance once the trees develop size and resistance to fire.

Summary

Management of pests in agriculture is extremely important to maintain our food supply. Plant and animal pathogens, insects, weeds, rodents and other pests have competed for human food since the beginning of recorded history.

Early methods of pest control were limited to hand picking, fire, use of salt or ashes, or just plain physical cutting or physically subduing plant or animal pests. Today we have excellent cultural treatments for crop production and chemicals that control weeds, insects, plant and animal diseases and varmints. Scientists have also significantly improved genetics in plants and animals to provide vigor, good nutrition, insect and disease resistance, drought tolerance and many other desirable features.

World expenditure exceeded more than $39 billion in 2007 and U.S. expenditures exceeded $12 billion in 2007 for pesticides. Greatest usage was for herbicides (40%) followed by insecticides, fungicides, and other pesticides.

Methods of pest control include chemical, mechanical, biological, and prescribed fire. Chemicals have been developed to control weeds, insects, plant and animal diseases, rodents and other pests of agriculture, urban areas, in and around commercial buildings, schools and homes. They have been developed for safe human, animal, and plant use and nontarget species. They are used at low dosages, must leave no harmful residue, and must be approved by the Environmental Protection Agency (EPA).

Biological control is a method to control pests with other organisms. It relies on predation, parasitism, herbivory, and other means. It can occur naturally and can be augmented by breeding and releasing natural enemies of the pest species.

Mechanical control is the use of handheld devices (hoes, sickles, shovels, axes, etc) or larger machinery such as disks, plows, blades, dozers, etc., mainly to remove weeds and woody plants.

Examples of pest management are given for corn in this text. Several types of herbicides can be used depending on weeds present, crop rotation, soil type, and environmental conditions. The first step in corn production is good cultural practices and selection of a good corn hybrid. The weed control practice may be cultivation only but many times involves herbicide use. Insecticides are available if insect problems reach critical numbers. Many disease problems can occur in corn and a resistant corn hybrid is desired.

All crops grown in the United States and worldwide may have insect, weed or plant disease problems that must be dealt with. These crops include soybeans, wheat, barley, grain sorghum, cotton, rice, alfalfa, corn, pastures and rangeland, forests and many other crops.

Chapter 13. Literature Cited

Grube A, Donaldson D, Kiely T, Wu I (2011). Pesticides Industry Sales and Usage: 2006 and 2007 Market Estimates. United States Environmental Protection Agency. Office of Chemical Safety and Pollution Prevention 33p. Retrieved from https://www.epa.gov/sites/production/files/2015-10/documents/market_estimate 2007.pdf. Accessed 20 October 2016.

Buchanan GA (1992). Trends in Weed Control Methods pages 47-72 in McWharter CG and Abernathy JR eds. Weeds of Cotton: Characterization and Control. The Cotton Foundation Reference Book Series, Publisher. Memphis, Tennessee.

Anonymous (2016a). Biological Pest Control. Wikipedia, the Free Encyclopedia. Retrieved from https://en.wikipedia.org/wiki/Biological_pest_control. Accessed 31 October 2016.

Ware GW and Whitacre DM (2004). *The Pesticide Book*, 6th Edition. Meister Pro Information Resources. Willoughby, Ohio. 487 p.

Jemison Jr JM and Bhowmik PC (2007). New England Guide to Weed Control in Field Corn. Bulletin #1124. The University of Maine Cooperative Extension. Retrieved from northerngraingrowers. org/wp_content/New-England-weed-mgmt-in-field-corn.pdf. Accessed 3 November 2016.

Anonymous (2016b). 2016 Corn Insect Control Recommendations. Retrieved from www.utcrops.com/cotton/cotton_insects/pub/ PB1768Corn.pdf. Accessed 3 November 2016.

Shurtheff MC, Edwards DI, Noel GR, Pedersen WL, Waite DG (2016). Diseases of Corn or Maize. (*Zea mays L.*). The American Phytopathological Society. Retrieved from www.apsnet.org/publications/commonnames/Pages/CornorMaize.aspx. Accessed 4 November 2016.

Anonymous (2016c). Diseases of Field Crops: Corn Diseases. Ministry of Agriculture, Food and Rural Affairs. Ontario, Canada. Retrieved from wwww.omafra.gov.on.ca/English/crops/pub811/14.com.htm. Accessed 4 November 2016.

Anonymous (2016d). Weed Control in Alfalfa. Kansas State University Research and Extension, Manhattan, Kansas. Retrieved from www.agronomy.k-state.edu/documents/weed-management/ppt-alfalfa-weed-control.pdf. Accessed 8 November 2016.

Anonymous (2016e). Facts and information on Alfalfa Pests. Department of Entomology, Kansas State University, Manhattan, Kansas. Retrieved from http://entomology.k-state.edu/extension/insect-information/crop-pests/alfalfa. Accessed 12 November 2016.

Hanson CH and Barnes DK (1975). Alfalfa, pages 136-147 in Heath ME, Metcalfe DS and Barnes RF eds. 3rd Edition Forage: The Science of Grassland Agriculture. The Iowa State University Press, Ames, Iowa.

Bovey RW (1987). Weed Control Problems, Approaches, and Opportunities in Rangeland, pages 59-88 in Foy CL ed. Review of Weed Science, Volume 3, 1987. Weed Science Society of America. Champaign, Illinois.

Bovey RW (2016). *Mesquite: History, Growth, Biology, Uses, and Management.* Texas A&M University Press, College Station, Texas. 259 p.

Casey M and Whittle P (2016). Invasive Insects Turn Forest into Wasteland. IPM in the South, Associated Press. Retrieved from https://pmsouth.com/2016/12/07/invasive-insects-turn-forests-into-wasteland.

FARM MACHINERY

Introduction

During the 18th century oxen and horses were used to pull crude wooden plows for farming. Sowing of crop seed was done by hand. Cultivation and weed control were done by hoeing. Hay and grain were cut by the sickle and grain and seed crops were harvested by flailing (Anonymous 2014).

Major agriculture events were the invention of the cotton gin by Eli Whitney in 1793, the first cast iron plow by Charles Newbold in 1797, and Thomas Moore of Maryland invented the icebox refrigerator in 1801 (Anonymous 2014).

The McCormick reaper was patented in 1834 and John Lane manufactured plows faced with steel saw blades. John Deere and Leonard Andrus manufactured steel plows and the threshing machine was patented in 1837.

In 1841 the practical grain drill was patented and in the mid and late 1800s the first grain elevator, irrigation, commercial fertilizer, windmills, horse drawn cultivators, steam and gasoline tractors, barbed wire, and a host of other improvements were made in farming (Anonymous 2014).

Hybridized corn was produced in 1881 and by 1890 agriculture became mechanized and commercial.

By 1920 to 1940 farm production grew slowly and expanded because of mechanized power. One farmer by 1940 could supply food for nearly 11 other humans. In the 1940s tractors slowly replaced horses for power and by 1950 one farmer could produce food for 16 people. By 1960 one farmer produced food for 26 people. In the 1970s no-tillage became popular and one farmer could feed 48 people. By 1990 one U.S. farmer fed over 100 people (Anonymous 2014).

Organic chemicals became available for insect, weed, and plant disease control after World War II in major U.S. crops and in the late 1990s the first weed and insect biotech crops of soybeans and cotton became available. Today Roundup ready corn, cotton, and soybeans are available (Anonymous 2014).

New technology includes the use of drones to scout fields for insects, weeds, and disease of crops and to apply other technology. Farm machinery is largely computerized to apply precise chemical and fertilizer applications, guide tractors, and monitor planting. Precision agriculture measures and monitors soil type, weed infestation, and insect problems to apply only needed and legal amounts of chemical at various locations in the field.

Early History

Progress in farming has been great and necessary to feed a large growing world population (9 billion by 2050). We also have to remember where we have come from and why we are able to sustain huge numbers of humans due to our continual growth and knowledge from new technology and innovation in agriculture.

The history of agriculture in the United States covers the period from the first English settlers to the present day. The colonists brought many ideas and some equipment from Europe. After 1800 cotton became the chief crop in southern plantations and the main American export. The number of farms grew from 1.4 million in 1850 to 2.2 million in 2008 (Anonymous 2016a).

The first settlers in Plymouth Colony planted barley and peas from England, but their most important crop was Indian corn (maize). To fertilize they used two small herring or shad fish in the soil when the seed was planted.

Most farming in the 19th century was to produce food for the family. In the South they grew their own food but exported cotton, tobacco, and sugar to Europe. However, with the dramatic expansion of the railroad, the number of farms increased from 2 million in 1860 to 6 million in 1905 as did farm value. The railroads also brought thousands of farmers from Germany, Scandinavia, and Britain (Anonymous 2016a).

Diary of a Farm Boy

The diary of an early American boy, Noah Blake 1805 (Sloane 1958), indicated a workable homestead in New England that the Blake family established. With the help of Noah (born in 1770), the homestead was gradually changed to include the Blake house, barn, mill, mill pond, and mill wheel to grind corn, toll house, corn field, uncut timber, a cellar, and other improvements and amendments.

Tools of axes, hammers, and nails were used to make all kinds of toys, ladders, hinges, gates, wooden sleds, boxes, fences, stump pulling devices, lifting devices, and other wooden devices to please the eye or help with work projects. The splitting maul and saw also helped make many objects such as tables and furniture, feeders, cellars, and apple presses. The wooden devices constructed were only limited by the imagination of the maker.

Seasonal Farm Jobs

Different jobs occurred in different seasons (Sloane 1962). Many farm jobs were done based on phases of the moon and their names. The reason for cutting wood depended on the type of tree and what the sap was doing within the tree. The old-timers believed that the moon affected all liquids on earth, like the ocean tides, and thus affected sap flow in plants. They believed cutting a tree three hours before the new moon and another of the same kind six hours afterwards made a difference in wood soundness.

The moon was also a timepiece and calendar for travel at night. Travel by full moon one could orient themselves better.

Each month of the year had jobs to be done on the farm (Sloane 1962). In New England, March and April were times to collect sap from the maple tree. The art was discovered by the American Indians as it didn't take the white American long to acquire a taste for sugar. The industry is alive and well in modern times and about 37 percent of the maple sugar comes from New England. March was also a time to lay up new-fuel wood for the hearth.

April was a time to clean, pick up stones and build stone fences and pull stumps. There were about five hundred kinds of stump-pullers on the market in the 1800s and it was a common practice. May was a time to plow the fields and plant corn and make brooms. Poles were most efficient if cast in May. At the time the bark on ash and hickory is high in oil making the bark easier to remove.

June was a popular season for weddings, growing crops, making tea, peppermint, and gathering herbs to dry. July was known as a season of weeds, hay making, and making cakes. August was a month of fence building.

September was hunting season, whether with the shotgun or trapping, and a time to fish. It was back to school season and a time to make pickles and harvest of crops. In October apple cider was made, apple butter, and beer from sorghums. The first proclamation of Thanksgiving for the United States was made in 1789 by President Washington for November. November was also a time to butcher livestock. Soap making also occurred made mostly of grease and fat from livestock and lye. Candle making season was also November to early December. November was also a time for cranberries. They were discovered by the Pilgrims. The Indians used them for food, medicine, and dye. Their Vitamin C content prevented scurvy. The first day of December was known as Sled Day wherever winter and snow were present. Farm sleds drawn by horses could be elaborate and fun. Wool was spun and the Christmas spirit abounded. January was for snowshoeing, ice making, and working in the farm shop repairing equipment. Farmers usually made their own tools and equipment. February was also used to close out the year and accounts and take inventory (Sloane 1962).

The early farmer made many contributions to tools, equipment, and techniques still used today. They were very inventive and productive and worked long hours each day to care for livestock, gardens, and crops. Today most of this is done for us and we just purchase it at the grocery or hardware store.

The Good Old Days

In a book edited by Kim and Janice Tate (various authors 2000) there are numerous stories about the "Good Old Days" working on the farm roughly 100 years ago. The theme is that everyone, even children, had chores to do. It was hard work but enjoyable work that usually required many hours of toil in the hot sun or cold winter. However, there were times of enrichment in which most folks felt they accomplished something very useful. Many of the same tasks as explained above had to be done to provide the necessities of life. The good old days were nostalgic but most of us would not like to go back to the endless toil and lack of modern material goods readily available to us today.

Farm Tractors

The history of the American farm tractor is intertwined with the history of American farming (Leffingwell 1991). The author remembers as a boy the large steam engines that were too large and cumbersome that soon were replaced by the small general purpose row crop tractor. Fordson, John Deere, and International Harvester Farmall were powered by gasoline engines.

The development of the gasoline engine tractors were based on the preceding 40 years of steam-tractor technology (Leffingwell 1991). At that time diesel engine power was also being investigated. Historians suggest that the earliest gasoline tractor engines resembled the steam engine because the familiar shape of the steam engine wouldn't frighten the horses or draft animals still in use (Leffingwell 1991).

The recommended power tools and blacksmith equipment to care for an early gas farm tractor were the drill press, smithing oven, vice, anvil, pliers, and hammer. Tractors during the first fifteen years of gas power grew larger as steamers had done. Their cost was substantial because of research and development. Only the largest farmers could afford them and they wanted large equipment to farm large acreages. By 1910, although there were over 1 million farms, only about 20 percent were over 500 acres in size (Leffingwell 1991).

The first World War needed horses and mules so farmers back home had to rely on tractors to get their work done, but most were unreliable. The Waterloo Boy Model N 12-25 hp passed the first test in 1920 as a result of a new law in the Nebraska House in 1919. To sell a tractor in Nebraska a firm had to have one tested in various prescribed ways. The Waterloo Boy passed the first time.

Fordson (Ford's tractor) had a big impact because they could produce the machine for less than their competitors, but Ford's expensive labor and extensive factories caused him to dramatically redefine the tractor business. In 1920 there were 186 tractor companies in the U.S. but few were left in 1928. International Harvester Company (IHC) beat Ford at his own price war and Ford quit in 1928. International Harvester Farmall and the John Deere Model D profited from Ford's innovations (Leffingwell 1991).

By the mid-1920s great improvements were made in farm tractors as a result of good engineering and farm participation (Leffingwell 1991).

Tractor power was rated in horsepower from the earliest days by pull of the drawbar and by power generated off the pulley wheel which was used for operating threshing machines, log sawing, and other tasks. The addition of the power take-off - a driveshaft running out of the rear of the tractor - powered implements towed behind it.

The diesel engine was perfected by German engineer Rudolf Diesel. In 1931 Caterpillar introduced its diesel tractor in Model 65. A small gasoline engine exhaust warmed the diesel cylinders and aided in the starting of the diesel engine.

The worldwide depression in the early and mid-1930s caused farm foreclosures and eliminated many more tractor manufacturers. By the late 1930s the number of tractors on farms more than doubled to well over a million. Tractor design and function improved and by 1939 Henry Ford returned to business with a new name attached to his tractors called the Irishman Harry Ferguson. It featured a three-point hitch which enabled the tractor's ability to plow while reducing the tractor's weight on it. The patented Ferguson System was so significant that other tractor manufacturers copied it.

The need for steel and rubber during World War II removed many obsolete tractors and machinery from U.S. farms. The steel was made into mortars and guns. By the end of the war, the U.S. Department of Commerce indicated tractor numbers were over 2.4 million and horse and mule numbers declined from nearly 17 million to about 12 million.

Many new developments flooded the early 1950s market. New hitching systems sped up the attachment of implements. Multispeed transmissions and gear-reduction systems were introduced. Liquified propane gas was introduced. Power steering and power implement lifts helped reduce farmers' work. Tractor seats were improved. Four-wheel drive tractors, dual wheels and tractors were built that flex in the middle for steering by the 1970s and horse power could be up to 500 or 600. Caterpillar even developed a rubber-tracked track layer in place of steel for some of their track layer vehicles (Leffingwell 1991).

Since the 1980s farms have become larger and many commercial. Many small farmers have retired or disappeared and the trend to large farms is a necessity to survive economically. One farmer fed 112 people in 1980, 129 in 1990, and 139 in 2000. Today one farmer feeds 155 people according to the American Farm Bureau Federation (Anonymous 2016b). Where will the number stabilize or will we be able to steadily increase farm productivity by new technology? The future will tell us.

Farm Machinery

The author remembers the machinery used during the 1930s, 1940s, and 1950s on the dryland farms of northern Idaho, Washington, and Oregon. On our farm and neighboring farms in northern Idaho, track layer D4 Caterpillar diesel-powered tractors were very common to pull heavy farm implements and combines. Track layer types also included the International Harvester tractors (TD-14) as well. Track layer tractors were very durable, stable, and able to pull implements on steep, rough soil. Many wheeled-tractors (gasoline-powered) were also used. On our farm we mainly used wheeled-tractors for economy. Wheeled-tractors were easier to drive from one field or farm to another and our land was not steep. We rated the tractor by the number of plows it could pull, and it was usually 3-4-plow category or approximately at 40-50 hp.

Our first tractor was an early 1930s Case which I recall as very durable and easy to drive. I was 10 years old in 1944 when I first had the opportunity to drive the tractor. We later also had the M-Farmall by International Harvester around 1940 and in the 1940s and 1950s a Minneapolis Moline, Model U, Model D John Deere, and Massey Harris (all gasoline-wheeled type tractors). Our neighbor had a wheeled type Fordson. My cousin still has a 1929 McCormick Deering (later International Harvester), a 1940 Farmall, and the early 1950s Massey Harris once owned by my father. Machinery was sometimes carried from place to place by tractor, trucks, or trailers.

We used the pulley wheel at the side of the engine to drive hay blowers and choppers, log sawing, and cutting firewood. The power take-off - a drive shaft running out of the rear of the tractor - drove sickle type mowing machines and other implements.

Our farm implements consisted of plows with 3-4 steel plows, disks, harrows, rod-weeders, sickle cutter mowers, hay rakes, hay choppers and blowers, small tractor-drawn combines, and drills to seed wheat, oats, barley, clovers, and peas. Combines and hay choppers were later self-propelled.

In row crop areas equipment was similar to dryland farming areas but wheel spacing on tractors were such as to be compatible in row crops as well as other pulled implements.

We didn't have the luxury of hydraulic lifts or power equipment but depended on a drawbar in which implements were attached. Some equipment like hay loaders attached to an M-Farmall tractor had hydraulics to pick up and stack hay as well as some other haying equipment and implements.

Today farm machinery is very sophisticated and specific. Hydraulics and quick attachments make operations much easier but costly. The ability to monitor all operations by sitting in the seat of an air-conditioned cab and making adjustments was only a dream of the pre- and post-World War II farmers. Today farming implements operate on the same principles as in the 1940s but everything is larger and faster. Limited till and no-till farming, as discussed in Chapter 4, in agronomic crops reduces soil disturbance and erosion, equipment needs, time in the field, labor, and costs. The use of genetically-modified crops ensures nutritional quality while reducing pest control costs. Driverless tractors and equipment increase safety, speed of operation, and hopefully reduce mistakes of operation. Future developments will be interesting.

Summary

During the 18th century, oxen and horses were used to pull crude wooden plows for farming in the United States. Sowing crop seed was done by hand, Cultivation and weeding were done by hand or hand-hoeing. Hay and grain crops were cut and harvested by the sickle and scythe. Seed and grain crops were harvested by flailing.

Major developments in agriculture were many; but invention of the cotton gin by Eli Whitney in 1793, the first cast-iron plow by

Charles Newbold in 1797, and the icebox refrigeration by Thomas Moore in 1801 were significant. The McCormick reaper patented by John Lane in 1834 was another milestone.

Other inventions in the 1800s included the grain drill, grain elevators, irrigation systems, commercial fertilizer, windmills, cultivators, barbed wire, steam and gasoline driven tractors. Hybrid corn was produced in 1881 and agriculture became mechanized and commercial.

The history of agriculture in the United States covered the period from the first English settlers to the present day. They planted barley and peas from England, but their most important crop was corn (maize) acquired from the Indians.

Most farming in the 19th century was to produce food for the family. In the south they grew food but exported cotton, tobacco, and sugar to Europe. Today one American farmer can produce enough food to feed over 155 people. Progress from the early pioneers made this possible by developing homesteads and clearing land for crops. The history of the American farmer is intertwined with the history of tractor development. Steam engines that were large and cumbersome were replaced by smaller gasoline-powered row-crop tractors. Fordson, John Deere, and International Harvester were early players. The early tractors of the 1900s were unreliable, but by the mid-1920s great improvements were made as a result of good engineering and farmer participation.

The diesel engine was perfected by German engineer Rudolf Diesel. In 1931 Caterpillar introduced its diesel tractor in Model 65. Diesel-powered wheeled and track layer tractors are now commonplace on U.S. ranches and farms. The need for steel and rubber during World War II removed many obsolete tractors and machinery from U.S. farms. The steel was made into mortars and guns. By war's end tractor numbers were 2.4 million and horse and mule numbers declined from 17 million to about 12 million. Today farm machinery is very sophisticated, specific, and sometimes large. Hydraulics and quick attachments make easier operation. High technology machinery and high technology crops are here to stay.

Chapter 14. Literature Cited

Anonymous (2014). Historical Timeline - Farm Machinery and Technology. Retrieved from https://www.agclassroom.org/gan/timeline/farm_tech.htm. Accessed 28 November 2016. Based on National Institute of Food and Agriculture (NIFA). United States Department of Agriculture (USDA) under Agreement 819.

Leffingwell R (1991). *The American Farm Tractor: A History of the Classic Tractor.* Motorbooks International Publishers and Wholesalers, Osceola, Wisconsin. 192 p.

Sloane E (1958). *The Seasons of America Past.* Dover Publications, Inc. Mineola, New York. 150 p.

Sloane E (1962). *Diary of an Early American Boy: Noah Blake 1805.* Dover Publications, Inc. Mineola, New York. 108 p.

Various authors (2000). *Good Old Days Remembers Working on the Farm.* Tate K and J. eds. House of White Birches, Berne, Indiana. 160 p.

Anonymous (2012). *The Complete Guide to Building Classic Barns, Fences, Storage Sheds, Animal Pens, Outbuildings, Greenhouses, Farm Equipment and Tools.* Atlantic Publishing Group, Inc., Ocala, Florida. 192 p.

Anonymous (2016a). History of Agriculture in the United States. From Wikipedia, the free encyclopedia. Retrieved from https://en.wikipedia.org/wiki/History_of_agriculture _in_the_United States. Accessed 30 November 2016.

Anonymous (2016b). Briefing on the Status of Rural America. U.S. Department of Agriculture Economic Research Service, Washington D.C. Retrieved from www.usda.gov/documents/ Briefing_on_the_status_of_Rural_America_Low_Res_Cover_up- date_map.pdf. Accessed 3 December 2016.

PRECISION AGRICULTURE AND NEW TECHNOLOGY

Introduction

Agriculture and agriculture production has come a long way since the early days described in Chapter 14. Agriculture production takes many forms but it takes proper and good equipment to be successful. In Chapter 14 we discussed the evolution of the tractor and how it influenced farming mechanization and evolution of farm implements.

On large farms huge tractors and many times specialized equipment are needed in production of new or improved crops and cropping systems. Speed of operation and application of multi chemicals and/or tasks can be done in one pass over the field in order to cover large areas in a short time period.

In planting winter or spring wheat in the Great Plains and Pacific Northwest U.S.A., a large tractor can pull seeding equipment to properly place and space small grain seed in the soil while applying fertilizer and other needed chemicals.

The John Deere 1860 no-till air drill is designed to open the soil, plant the seed at the proper depth, and close the soil again in a single pass. The drill adjusts to soil types and depth (Carroll 2013).

In many parts of the world most crop planting, growing, and harvesting is still done by hand with or without the use of tools. Although minimum or no-till farming may be used as described above in the U.S.A, and many areas of the world, plowing, disking, and harrowing are still commonly practiced before seeding and after harvest. For these operations the tractor has become a vital farm implement even in countries less developed. Once these operations are done and the soil smoothed, seeding can be done. Air-conditioned and dust-proof cabs on tractors are desired.

Where possible minimum till or no-till systems help reduce wind and soil erosion, reduce fuel and equipment use and cost. Some farmers like the conventional plow-disk-harrow method because they have the equipment and feel no-till is not as productive as conventional methods.

Row-crop rippers, designed for various field applications including ripping and bedding, are used in some crops such as cotton. The fields are bedded in the fall and can be planted on raised beds in the early spring in areas of high soil moisture.

All types of special equipment include tractors with wide and rubber tracks to prevent soil compaction, tricycle-type wheels for row crop work, crop sprayers, 4x4 tractors, compact tractors, and tractor pulling as a sport. Machines to pull carrots, dig potatoes, collect apples, and harvest lettuce and cabbage are available.

Threshing of grain in the old days was all done by hand but today self-propelled combines harvest small grains, corn, cotton, blackberries, and many other crops. Hay is cut and conditioned, raked, bailed, chopped, or put up loose depending on the desires of the farmer by use of specialized machines.

Precision Agriculture

Precision agriculture or satellite farming is a farming management concept based on observing, measuring, and responding to inter- and intra-field variability in crops. It is possible because of GPS and GNSS. Maps are created to indicate differences in crop yield, terrain features/topography, soil organic matter content, moisture levels, pH, electrical conductivity (EC), nutrient levels, etc. Similar data is collected by crop yield monitors mounted on GPS-equipped combine harvesters or other vehicle mounted sensors that measure these variables. This data is used by variable rate technology (VRT) on seeders, fertilizer spreaders, pesticide applicators, and similar equipment to optimally distribute seed, fertilizer, herbicides, insecticides or fungicides and similar materials needed in optimal and economical growth of crops (Anonymous 2016a).

Geolocating a field enables the farmer to analyze data after delineating the field using an in-vehicle GPS receiver on a tractor driven around the field or by a base map derived from aerial or satellite imagery. The concept of precision agriculture first appeared in the 1980s by researchers at the University of Minnesota that varied lime application in fields. The grid sampling practice also appeared at that time by applying a fixed grid of one sample per hectare. By the end of the 1980s this technique was used to drive input maps for fertilizers and pH corrections. The use of yield sensors developed from new techniques combined with the advent of GPS receivers has been gaining ground ever since. Today such systems are used on several million hectares in the U.S. and worldwide. This practice allows farmers to vary fertilizer and pesticide rates across a field at the correct time in correct amounts to maximize crop yield, save costs, and be more environmentally friendly (Anonymous 2016a).

Other uses of global positioning systems (GPS) and GPS-computer guided tractors and harvesters include electro-magnetic soil mapping, soil sample collection, crop yield data collection, aerial imagery of crop or soil color index maps, soil types, soil characteristics, drainage level, and potential yields (Soil Science Society of America [SSSA] 2015).

Precision agriculture has also been enabled by affordable unmanned aerial vehicles like the DJI Phantom that costs under $1000 and can be operated by novice pilots. These drones can be equipped with hyperspectral or RGB cameras to capture many field images that can be photogrammetric methods to create othrophotos of NDVI maps (Anonymous 2016a).

Agricultural Innovations Cost and Acceptance

Sunding and Zilberman (2000) indicated that economic forces as well as scientific knowledge influence the form of innovations that are generated and adopted in various agriculture locations. For example, chemical and biotechnological innovations are related to problems of public acceptance and environmental concerns. Mechanical and technological innovations may negatively affect labor and lead to farm consolidation. They may include new tractors, machinery and harvesters, new seed cultivators and varieties, chemical innovations (fertilizers and pesticides), and new management practices that change farming practices. These new practices may or may not be economical to use or accepted in many cases.

It may take several years before a farmer will accept a new practice, machine, or chemical because of cost or reluctance to change. Some changes in agriculture are slow because it has never been done before. The change may also be too costly in dollars, environmental, or human concerns.

Many things can change the grower's perception of adopting new ideas and practices, much of which depends on a stable seller's market, available labor, resources including land and favorable climate.

Labor shortages can also induce labor saving technologies.

Global Forum and Innovations in Agriculture

More than 80 of the world's most exciting innovations for sustainable agriculture were presented at the GFIA 2016 forum (Anonymous 2016b). Some include 1) wastewater recycling system to revolutionize dairy farms, 2) a greenhouse that waters itself in the desert, 3) high-tech bee backpackers to secure the future of food, 4) biologicals for improved pest, disease, pollination management, 5) early mastitis detection in dairy cows, 6) a self-sufficient greenhouse using solar, wind, geothermal and biomass methods, 7) democratized use of professional drone mapping software, 8) using drones and cropping culture, 9) saving time and money using agricultural robotics, and 10) using satellite data for decision-making in crops just to name a few. The new forum will take place on March 20-21, 2017 at Abu Dhabi National Exhibition Centre - UAE.

This global forum indicates that agricultural innovations are not limited to farm tractors and machinery but include all the steps in agriculture production and the food supply. All the chapters we have discussed up to now involve our food supply, whether packing or processing of foods, plant science, animal science, land resources, or water supply.

Robots, sensors, automation, and engineering changes according to Policy Horizons Canada with revolutionary agriculture (Zappa 2014). This includes sensors to trace and diagnose the status of crops, livestock and farm machinery, genetic tailoring of meat in the laboratory, and robots and microrobots to check crops at the plant level. Engineering will be involved in synthetic biology in making fuels, byproducts from organic chemistry, and smart devices (Zappa 2014).

The use of sensors, automation, robots, and engineering is almost limitless and is limited only by the imagination and ideas of the innovators and sponsors. New inventions and innovation have been a characteristic of the civilized world and our early farmers and ranchers have shown this unlimited boundry since the Pilgrims landed in Plymouth Rock, U.S.A.

Seager (2014) lists six innovations revolutionizing farming including: 1) dairy hubs, 2) fertilizer deep placement, 3) mobile apps, 4) high-roofed greenhouses, 5) new feeding systems, and 6) farm management software and training. The information is mainly for small farm holders and developing areas.

Buchanan (2016) points out the critical need for agricultural research because the world population is expected to exceed 9 billion before 2050 and our petroleum use is rapidly increasing on a finite supply.

Dr. Buchanan presents a list of 22 grand challenges for agriculture research and innovations as follows:

1. Improve soil quality
2. Improve agricultural energy efficiency
3. Use of nitrogen fixation in non-legume plants
4. Enable C_3 plants to utilize C_4 photosynthetic pathway
5. Efficient conversion of cellulose, hemicellulose, and lignocellulose to a more useable energy source
6. Enhance water efficiency in plants
7. Enhance more efficient nutrient use in plants
8. Improve pest resistance in plants and livestock
9. Eliminate livestock diseases
10. Improve nutrition in all food animals
11. Develop hybrid vigor in crops where possible

12. Incorporate apomixis in crop plants (production of seed without fertilization by the male gamete in pollen) (Hybrid-seed production can be circumvented by apomixis)
13. Improve plant response to elevated CO_2 levels
14. Improve energy efficiency of plants
15. Develop commodities of increased health benefits
16. Develop stress tolerance in plants
17. Delay plant senescence (enhanced productivity)
18. Prevent or capture animal waste
19. Use genomics and biotechnology to improve food animal quality
20. Seek new innovations
21. Develop new water harvesting methods
22. Improve food storage

If Dr. Buchannan's list was supported and funded, much could be done to properly feed the rapidly increasing world population. Agricultural innovations and technology have been a wonderful and productive focus for many years. New ideas and technology are being discovered and made available in agriculture and food production every day.

Compact Farms

Most of us think of farming on a large scale, whether family or commercial farms; however, small farms have existed for a long time and, as indicated in Chapter 5, small farms exist in the United States and around the world.

Volk (2017) describes 15 proven plans for market farms on 5 acres or less in the U.S. or Canada that are productive and profitable. The author concludes that maybe it is time for a flood of compact farms to grow food in North America as a renewed technology for urban agriculture.

The Vertical Farm

Despommier (2010) describes a compelling method of growing crops in multistory buildings using hydroponic and aeroponic farming methodologies. Such crops as strawberries, tomatoes, peppers, cucumbers, herbs, and a wide variety of special crops, as well as fish and poultry, could be grown in vertical farms. This would allow food production in the urban setting, where needed, significantly reducing food transport and storage facilities. Farmland in need of restoration could be released or used for other purposes. Systems should be developed to purify and utilize human, animal, and plant waste and residue and grow human and livestock food. The use, construction, cost, and operation should be evaluated by professionals and pilot studies for this concept. Vertical farming should be seriously considered in view of the projected world population, environmental concerns, and loss of agricultural lands to other uses.

Summary

Agriculture and agriculture production has come a long way since the first pilgrims landed on Plymouth Rock, U.S.A. Much of the progress in agriculture can be attributed to development of mechanization and the evolution of tractors replacing draft animals. Equally as important is the improvement of livestock and crops by selection and improved genetics. Minimum till and no-till cropping systems have reduced wind and water erosion, reduced fuel and equipment use and cost. Specialized equipment has been developed to farm large acreages and do special jobs in row crops, pulling carrots, digging potatoes, apple collection, and harvesting lettuce and cabbage.

Threshing grain in early times was done by hand; but today self-propelled combines harvest small grains, corn, cotton, blackberries, and many others crops. Hay is cut, conditioned, raked, bailed, chopped, or put up loose by modern machinery.

Precision agriculture or satellite farming is a farming management concept based on observing, measuring, and responding to inter- and intra-field variability in crops. It is possible because of GPS and GNSS. Maps are created to indicate differences in crop yield, terrain features, organic matter content, pH, EC and nutrient levels, etc.; and the necessary adjustments are made to obtain maximum crop yield and quality from data collected.

New innovations are frequently made for small and large farmers alike in equipment, technique, and chemicals. Cost, farmer acceptance, and environmental concerns may prevent adoption of new practices. However, some of these practices may be adopted because of improved benefits, reduced labor costs, or they are more environmentally friendly. Some new innovations have been listed as well as agricultural research and funding needs to promote, expand, and increase our food supply.

Chapter 15. Literature Cited

Carroll J (2013). *The Illustrated Encyclopedia of Tractors and Farm Machinery.* Lorenz Books - Anness Publishing Ltd, Blady Road, Wigton, Leicestershire LE18 4SE. 256 p.

Anonymous (2016a). Precision Agriculture. Retrieved from Wikipedia, the Free Encyclopedia. https://en.wikipedia.org/wiki/Precision_Agriculture. Accessed 22 December 2016.

Soil Science Society of America (2015). What is "Precision Agriculture" and Why is it Important? Retrieved from https:// soilsmatter.wordpress.com/2015/02/07/what-is-precision-agriculture-and-why-is-it-important/ Accessed 22 December 2016.

Sunding D and Zilberman D (2000). The Agriculture Innovation Process: Research and Technology Adoption in a Changing Agricultural Sector (For the Handbook of Agricultural Economics). Retrieved from are.berkeley.edu/~zilber11/ARE241/fall 2005/ innovationchptr.pdf. Accessed 27 December 2016.

Anonymous (2016b). Innovation Programme at GFIA 2016. Global Forum from Innovations in Agriculture. Retrieved from www.innovationinagriculture.com/Conference/Innovations. Accessed 29 December 2016.

Zappa M (2014). 15 Emerging Agricultural Technologies That Will Change the World Policy Horizons. Canada. Retrieved from www.businessinsider.com/15-emerging-agriculture-tehnologies-2014-4. Accessed 29 December 2016.

Seager C (2014). Six Innovations Revolutionizing Farming. The Guardian. Retrieved from https://www.theguardian.com/global-development-professionals-network/2014/jul/08/top-six-innovations.sm. Accessed 29 December 2016.

346

Buchanan GA (2016). *Feeding the World: Agricultural Research in the Twenty-First Century.* First Edition. Texas A&M AgriLife Research and Extension Service Series. Texas A&M University Press. College Station. 322 p.

Volk J (2017). *Compact Farms.* Storey Publishing. North Adams, MA. 227 p.

Despommier D (2010). *The Vertical Farm: Feeding the World in the 21st.* Picador, New York, NY. 311 p.

POLLUTION AND POLLUTION MANAGEMENT

Introduction

No discussion of farming and animal and plant production is complete without mention of possible pollution with agricultural chemicals, livestock wastes, and general farming practices.

Although most people are in business to make a profit from their agricultural enterprise, they need to maintain farm buildings, machinery, livestock, land and water resources in good condition. In doing so, they are good stewards of their resources and are environmentally responsible.

Maintaining land and water sources is particularly important because it not only affects the landowner but may have implications for nearby neighbors and the entire community and may for generations to come. Each responsible rancher and farmer sets the example for others by doing good work and keeping up to date on agricultural chemicals, farm machinery, conservation, crop and livestock methods, and new technology.

In the following pages we will discuss the importance of pesticide and herbicide use, fertilizer application, animal wastes, insect and weed problems, soil compaction and erosion, tillage, and protection of water resources. We will also discuss any environmental problems encountered in the ranching and farming business and how to correct or remediate those problems.

Pesticide and Herbicide Use

Pesticide and herbicide use are generally safe to use if the directions on the pesticide label are followed. It contains all the information needed to apply an agricultural chemical. It contains information as a result of extensive years of research and development, efficacy on the target pest, and safety precautions to be met (Ware and Whitacre 2004).

Chemical and pharmaceutical companies work hard to discover and develop medicines and agricultural chemicals that are effective on the target pest while friendly to the environment and nontarget species.

Agricultural chemicals used today are necessary in agriculture to produce the abundant crop and livestock products we need to survive and enjoy and those we export to other countries. Agricultural chemicals are strictly approved, registered, and administered by the U.S. Environmental Protection Agency (EPA). States also have agricultural chemical regulations that are in line with EPA but can be more restrictive than the EPA.

Market basket surveys of food rarely indicate pesticide or herbicide residues in prepared or processed food. If found, a pesticide would probably be so minute in content that it would be biologically insignificant. Today instrumentation to detect agricultural chemicals are so sophisticated that they can detect levels in the low parts per trillion (PPT) that are infinitesimal. That doesn't mean that they are not of concern, but are likely not very significant.

On our small grains and cattle farm in northern Idaho soon after World War II, we had a Canada thistle weed problem and hairy vetch that severely competed with the growth of wheat, barley, and oats. When 2,4-D (herbicide) became available, we used it in small amounts as a spray in water in the crops to control these weeds. It controlled the weeds without injury to the crops. It could be sprayed by ground or aerial sprayers and was a tremendous boost to crop production.

Some people refer to the weeds as pollution rather than an agricultural chemicals. The same may be said for insect pests and disease organisms on forage and crop plants and livestock.

However, should an agricultural chemical of high toxicity cause a serious problem or accident, a toll-free long distance number can be called CHEMTREC (800-424-9300) HOTLINE. Another number to call is the National Pesticide Information Center Hotline (800-858-7378) (Ware and Whitacre 2004). The third service is for human poisoning cases. It is the nearest Poison Control Center.

There are many examples, plus the one given above, that indicate the usefulness and safety of a pesticide or herbicide. The herbicide 2,4-D was developed in the 1940's and is still widely used in the United States and worldwide. It is short lived in the environment and soils and is not a problem in water sources.

Fertilizer Use

In Chapter 2 we discussed in detail fertilizers and fertilizer use. We described the nutrients and micronutrients necessary for plant growth and crop production. Modern farming and horticulture would not be very productive without applications of nutrients needed for growth. Livestock also depend upon good forage and supplemental feed of good nutritional quality.

If nutrients are taken out of the soil with crop and animal production over long periods of time, then those nutrients, such as nitrogen and phosphorus and others, need to be replaced in order to maintain productivity.

In some cases overfertilization may occur or nutrients are carried to unwanted areas by runoff water or sediment (soil) erosion. These nutrients can end up in streams, rivers, and lakes and cause accelerated eutrophication (algal blooms and other effects from excess nitrogen and phosphorus).

The symptoms of eutrophication are increased biomass of phytoplankton, attached algae and macrophytes. It increases aquatic plants, affects water quality, clogs waterways, impedes water navigation, causes noxious odor from decaying vegetation, and kills fish.

Eutrophication is not a problem everywhere but just in isolated cases. Agriculture is a big contributor to the eutrophication problem and eutrophication can be closely associated with rainfall-runoff events from treated areas. Phosphorus runoff losses are directly associated with soil erosion (sediment).

There are no easy solutions to eutrophication because it is a natural process.

Contaminated runoff water only enhances it. Applying fertilizers in the proper amount and method when plants can fully utilize it reduces potential pollution. Planting cover crops that keep nutrients out of water by recycling excess nitrogen and phosphorus helps. It also reduces potential soil erosion. Buffer zones of grasses, shrubs, or trees around fields and waste areas, especially those that border water bodies, can help absorb or filter out excess nutrients before reaching a stream or lake.

Conservation tillage (reduced tillage or no till) reduces soil erosion and soil compaction, enhances organic matter content, and reduces runoff water. Conservation tillage has some controversy but is now widely practiced on United States farmland and worldwide (Anonymous 2016).

Another way to prevent nutrient loading is to reduce or restrict runoff water from agricultural fields that may contain surplus nutrients that may drain into adjacent streams and lakes.

Lakes suffering from eutrophic effects may have lake bottom sediments of phosphorus enriched particles. They can be dredged or chemically treated to immobilize the phosphorus but lake restoration is expensive.

Animal Wastes

Livestock and poultry manure is rich in nitrogen and phosphorus like some fertilizers. Keeping animals and their waste out of streams, rivers, and lakes keeps nitrogen and phosphorus out of the water and helps restore stream banks. Using the same techniques as with commercial fertilizers will abate the unwanted residues and movement of excess animal manure.

On the other hand, animal manure can best be disposed of by using it as a fertilizer in the proper amount and proper timing for plant growth. It adds valuable organic matter and nutrients to the soil or growth media and is a perfect way to dispose of it. Livestock and poultry producers could dispose of excess manure by selling it to other growers and gardeners.

Some farmers are pelletizing and selling farm wastes for a second income to people who want sterilized manure for organically grown crops and produce. Each year farm animals in the United States produce over 335 million tons of manure (Anonymous 2014).

Storing manure on the farm until further use can be done safely with the proper location and facility.

Soil Erosion

It has been stated that it takes 1000 years to produce one inch of soil, so soil is a valuable resource. It also takes a very short time to lose that inch of soil with mismanagement or acts of nature. Without fertile soil, growing plants and livestock becomes difficult so every attempt should be made to conserve what nature has given us. As we learned in Chapter 2, all soils are derived from rock and most nutrients so taking care of land and water resources is necessary.

Land in permanent pasture, rangeland, and/or forest is usually stable and resistant to wind and water erosion. However, if overgrazed or denuded of vegetation in some way, soil can be displaced by forces of nature.

Farmers and ranchers take great pride in caring for their land because it must serve them well and should be passed onto future generations in good shape. Farmland can be used generation after generation if soil nutrients are added back in fertilizer, manure, or plowing under green cover crops. Wind and water erosion are slowed by leaving the soil surface rough over winter. Proper drainage of excess water is done and buffer zones of grass-vegetation are strategically located on slopes, field borders, or between fields that are cropped. Plowing (tilling) on the contour of a hill, if possible, may be better than straight up and down the slope. Adopting minimum till and no-till farming may be better than conventional farming. Levees or dams placed strategically on sloping fields can slow and alter water flow to slow soil erosion.

It may sometimes be better to put the land back in permanent vegetation and pasture rather than trying to farm marginal property. Loss of soil can be tragic and lead to abandonment of the land so every precaution and method should be applied to preserve and improve the property you own or lease. The Natural Resource Conservation Service, the Extension Service, state universities, and special consultants can help with farm and ranch planning to make best use of the land. Land quality varies and each ranch or farm needs to be evaluated as to its best and productive use based on soil type, length of growing season, climate, precipitation or supplemental irrigation, and crops adapted to the area.

Pests and Weeds

Some consider certain pests of crops pollutants such as weeds, insects, and animal and plant disease problems. Weeds, as well as insects, can destroy crops as well as injure farm animals. Certain plant and animal diseases can also cause great damage and reduce productivity. Overgrowth of aquatic weeds can render streams unnavigable, kill fish, and interfere with water flow. Weeds can compete with crops to absorb water and nutrients and light needed for crop growth and render the crop unproductive. Small animals and birds can feed on crops before harvest and destroy the seed and/or foliage. These pests have been discussed in Chapter 13.

Summary

Pollution of agricultural lands can occur by misuse of agricultural chemicals, livestock wastes, and general farming practices. Pesticides and herbicides are generally safe to use for humans, livestock, and the environment if the directions are carefully followed on the pesticide label. Agricultural chemicals, including fertilizers, are necessary in today's agriculture to feed a growing population. Eutrophication (algae blooms and other effects from excess nitrogen and phosphorus) can occur if fields are over-fertilized and contaminated runoff water ends up in streams or lakes. Eutrophication may cause accelerated biomass growth causing increased aquatic plants, clogged waterways, and noxious odor from decaying vegetation. Prevention of excessive eutrophication can be done by applying the proper amount of fertilizer, planting cover crops, preventing soil erosion, using conservation tillage, and restricting excessive runoff from agricultural fields.

Livestock and poultry manure are rich in nitrogen and phosphorus. Keeping animals and their waste out of streams and lakes using the same techniques as with commercial fertilizers will work for livestock waste. However, animal manure can best be disposed of by using it as a fertilizer in proper amounts and timing for plant growth. It adds valuable organic matter and nutrients to the soil. Some livestock growers sterilize and pelletize excess animal wastes and sell it to other growers and gardeners.

Soil erosion and soil movement is a natural ongoing process. Production of good soil for plant growth takes many generations but only a few seconds to sometimes lose it by acts of nature or human activity. For farming and ranching purposes, good soil and good soil conservation is required. Land in permanent grassland or forest is stable and resistant to wind and water erosion. On cultivated land, wind and water erosion are slowed by leaving the soil surface rough over winter. Conservation or no-till cultivation reduces wind and water erosion, increases water infiltration, reduces runoff, and preserves soil nutrients.

Chapter 16. Literature Cited

Water GW and Whitacre DM (2004). *The Pesticide Book*. 6th Edition. Meister Pro Information Resources. Willoughby, Ohio. 488 p.

Anonymous (2016). Nutrient Pollution. The Sources and Solutions: Agriculture. United States Environmental Protection Agency (EPA). Retrieved from https://www.epa.gov/nutrientpollution/sources-and-solutions-agriculture. Accessed 5 January 2017.

Anonymous (2014). Modern Farmer. What to Do with All of the Poo? Retrieved from modernfarmer.com/2014/08/manure-usa/. Accessed 5 January 2017.

MARKETING AND POLICIES

Introduction

In the United States, government policy and agricultural productivity were connected when President George Washington in 1799 established with Congress the National Agricultural Board. In 1862 President Abraham Lincoln signed an Act of Congress establishing the United States Department of Agriculture (USDA) which allowed farmers to find new ways to cultivate land and produce a safe and abundant food supply (Anonymous 2016a).

The USDA is the primary federal agency that promotes, regulates, and enforces government policy and the American farm industry. The primary mission is to implement policies approved by Congress every five years in what is known as the "Farm Bill." It authorizes federal funds for farm subsidies, food aid, nutrition programs, rural development, trade, farm credit regulations, conservation programs, market support, and so on (Anonymous 2016).

The U.S. Food and Drug Administration (FDA) has jurisdiction over how some foods are handled, prepared, and stored. The Environmental Protection Agency (EPA), we have discussed before, enforces air and water quality laws and application of farm chemicals. It also regulates the renewable fuel standard (RFS) which dictates how much biofuel must be used.

Other federal agencies provide funding for scientific research, renewable energy, transportation infrastructure, and more. Federal agencies that impact the economy, foreign affairs, and trade all affect agriculture (Anonymous 2016a).

At the state level, government agencies promote local agricultural products, provide food safety, inspection services, soil conservation, and environmental protection. States regulate production, transportation, processing, and marketing of commodities. Many state universities and colleges provide agricultural education, research, and extension services necessary for the survival of agriculture. At the local level, county and municipal governments promote agricultural education and regulate some farm operations, markets, community gardens, and food assistance programs (Anonymous 2016a).

Not everyone may agree with some governmental policies, especially some of the EPA's stringent regulations, but some regulation is needed to maintain essential products and services.

Bakst (2017), however, feels that markets and not government incentives and controls should inform farming decisions. Bakst (2017) pointed out that government policies drive up food prices and mentions the sugar program and the renewable fuel standard as examples. Bakst (2017) argues that subsidies are not necessary for farmers to succeed and that government regulations should be minimized with sound regulations. Obstacles to new agricultural innovations should be removed including taxpayer-funded research that discourages private research. Trade opportunities are lost when Congress subsidizes domestic U.S. agriculture industries and promotes special interests.

Tweeten and Amponsah (1996) explored a number of options for public policy to better serve small farms. They concluded that improved extension education and human resource development offer some of the most promising public policy opportunities to help small farmers.

Farmer's Market Policy

Hamilton (2006) assesses the current farmer's market policies found in the U.S. and uses this inventory to evaluate the effect of policies and investigates how they can be improved. He asked 1) what regulatory or policy obstacles do farmers markets face? 2) how have local markets overcome policy obstacles? and 3) what are federal, state, and local officials doing to facilitate farmer's markets? Hamilton (2006) has specific and detailed suggestions for improvement of farm markets at the federal, state and local level. The idea is to provide stable markets and good nutrition for all U.S. citizens. Hamilton (2006) lists several website organizations promoting farmers markets. A few include North American Farmers Direct Marketing Association (NAFDMA), Farmers Market Coalition (FMC), USDA Agricultural Marketing Service, Northeast-Midwest Institute, Community Food Security Coalition, Institute for Agriculture and Trade Policy (IATP), National Campaign for Sustainable Agriculture, and others.

Farmers Market Coalition

Farmers Market Coalition (FMC) is one of the organizations mentioned by Hamilton (2006). The FMC works to equip market managers and farmers with the tools necessary to run successful markets. Information is shared through webinars, newsletters, and their resource library. FMC gives farmers markets a voice in public policies that impact the food system. FMC ensures local farmer interests are understood and represented at the state and federal levels. FMC supports market managers, vendors, and customers and works closely with regional leaders to build state farmers market networks and associations (Anonymous 2006b).

American Farm Bureau Federation (AFBF)

The American Farm Bureau Federation, commonly referred to as the Farm Bureau, is a nonprofit and largest farm organization in the United States. The mission is to work with rural Americans to build strong, prosperous agricultural communities. AFBF is headquartered in Washington DC and includes all 50 states and Puerto Rico. The Farm Bureau started in 1911 by a Broome County extension agent. The effort was led by the U.S. Department of Agriculture and the Lackawanna Railroad. AFBF soon spread to other places (AFBF 2017).

In 1914 the Smith-Lever Act of 1914 was passed by Congress to share costs of county agent programs to provide information to farmers on improved methods of animal husbandry developed by agricultural colleges and agricultural experiment stations. It evolved into the modern day Cooperative Extension Service.

The AFBF was formally founded in 1919 in Chicago, Illinois, separate from government, for the purpose of making the business of farming more profitable and the community a better place to live. The Farm Bureau is an independent, nongovernmental, voluntary organization governed by and representing farm and ranch families to solve problems and formulate action to achieve educational improvement, economic opportunity, and social advancement. It is local, county, state, national, and international in scope and influence. It is the voice of agricultural producers at all levels (AFBF 2017).

Market and Government

Weingast (1995) stated that thriving markets require not only an appropriately designed economic system but a secure foundation and limited government that limits its ability to confiscate wealth.

This arose in the developed West, like England and the United States.

Markets and the USDA

Some important agencies of the USDA related to marketing and policies are Agricultural Marketing Service (AMS), Agricultural Research Service (ARS), Animal and Plant Health Inspection Service (APHIS), Center for Nutrition Policy and Promotion (CNPP), Economic Research Service (ERS), Farm Service Agency (FSA), Food and Nutrition Service (FNS), Food Safety and Inspection Service (FSIS), Foreign Agricultural Service (FAS), Forest Service (FS), Grain Inspection, Packers and Stockyards Administration (GIPSA), National Agricultural Library (NAL), National Agricultural Statistics Service (NASS), National Institute of Food and Agriculture (NIFA), Natural Resources Conservation Service (NRCS), Risk Management Agency (RMA), and Rural Development (RD) (Anonymous 2015).

There are many USDA offices also. One important office is the Office of Advocacy and Outreach (OAO) established by the 2008 Farm Bill to improve access to USDA programs that improve profitability and viability of small farms and ranches. Another is the Office of the Chief Scientist (OCS) to identify and prioritize agricultural research, education, and extension; and another is the Office of the General Counsel (OGC) that provides legal advice and service to all USDA programs and activities.

In the United States, Agricultural Marketing Service (AMS) USDA mentioned above has programs for cotton, dairy, fruit, vegetables, livestock, seed, poultry, and tobacco. This program tests, standardizes, grades, and markets new services and administers agreements and order for commodities for federal food programs. AMS also enforces certain federal laws and supports the Agricultural Market Resource Center at Iowa State

University and Penn State University. Other federal agencies are also involved in marketing and policies of food and agriculture.

World Trade Organization

The World Trade Organization (WTO) deals with global rules of trade between nations. It ensures trade flows as smoothly and freely as possible. The United States is a member of the WTO. The United States has free-trade agreements (FTAs) with about 20 countries.

Summary

In the United States, government policy and agriculture productivity were connected when President George Washington in 1799 established with Congress a National Agricultural Board. President Abraham Lincoln signed an Act of Congress to establish the United States Department of Agriculture in 1862 to find new ways to cultivate land and produce a safe, abundant food supply. The USDA is the primary federal agency to promote, regulate, and enforce farm policies implemented by Congress. The "Farm Bill" authorizes federal funds for farm subsidies, nutrition programs, rural development, trade, credit, conservation, markets, and so on. State governments promote and regulate agricultural products, food safety, inspection services, and conservation and environmental programs. Local governments provide education, farm operations, markets, and food assistance programs. Some individuals feel agriculture would be more successful with lower food prices if government involvement was minimized. Farm markets are very complex and many federal, state, and local governments, plus many farm organizations, promote good farming and marketing. Government is heavily involved and should provide a secure and sound economic system with limitations for agriculture markets to thrive.

Chapter 17. Literature Cited

Anonymous (2016a). The Role of Government in Agriculture. Iowa Public Television, Johnston, Iowa. Retrieved from site.iptv.org/mtom/classroom/module/14002/role-of-government-in-agriculture. Accessed 7 January 2017.

Bakst D (2017). 10 Guiding Principles for Agriculture Policy: A Free Market Vision. The Heritage Foundation. Retrieved from www.heritage.org/research/reports/2014/05/10-guiding-principles-for-agriculture-policy-a-free-market-vision.Accessed 7 January 2017.

Tweeten LG and Amponsah WA (1996). Alternatives for Small Farm Survival: Government Policies Versus the Free Market. Journal of Agricultural and Applied Economics, 28(1): 88-94.

Hamilton ND (2006). Farmers Market Policy: An Inventory of Federal, State and Local Examples. Drake University Agriculture Law Center, Des Moines, Iowa. Retrieved from https://www.pps.org/pdf/FarmersMarketPolicyPaperFINAL.pdf. Accessed 8 January 2017.

Anonymous (2016b). Farmers Market Coalition. Retrieved from https://farmersmarketcoalition.org. Accessed 9 January 2017.

American Farm Bureau Federation (AFBF) (2017). American Farm Bureau Federation, Washington DC. Retrieved from https://en.wikipedia.org/wiki/American_Farm_Bureau_Federation. Accessed 9 January 2017.

Weingast BR (1996). The Economic Role of Political Institutions: Market-Preserving Federalism and Economic Development. The Journal of Law, Economics and Organization, VII NI. Retrieved from web.stanford.edu/~jrodden/osio/weingast_mpf.pdf. Accessed 8 January 2017.

Anonymous (2015). USDA Agencies and Offices. Retrieved from https://www.usda.gov/wps/portal/unda/usdahome?navtype=MA&navid=AGENCIES_OFFICES. Accessed 10 January 2017.

Anonymous (2017). Agricultural Marketing. From Wikipedia, the free encyclopedia. Retrieved from https://en.wikipedia.org/wiki/Agricultural_Marketing. Accessed 10 January 2017.

CHAPTER 18

TRAINING NEEDS AND EDUCATION

Introduction

Many of the great farmers in the United States and worldwide have no formal education but learned from doing or from their parents or neighbors. Many of our U.S. farmers are seniors because getting into farming is costly, a long-term affair, and financially risky. Young people either inherit the land, machinery, and buildings or have a relentless desire to farm.

With narrow profit margin and high technology required, farming isn't what it was 50 years ago and the modern farmer to be successful needs much assistance and knowledge. They can get that assistance and knowledge in many ways by attending farm and ranch meetings, short courses in farming from local schools and colleges, or formal training at a nearby university. There is also much information online, radio, television, and local newspapers from the extension service, state university, USDA, or industry people.

I once asked a farmer that farmed a large area if his college major was agriculture, in which he responded that he knew how to farm but lacked business knowledge. He majored in business.

A wise choice for him since he was at least a third generation farmer.

I will attempt to list the training and knowledge needed for agronomic, horticulture, tree crops, and livestock production and other agriculture situations.

Needs for Crops and Horticulture Farming

First the upstart farmer should have a knowledge of soils, how they are classified and should be managed relative to water needs, drainage requirements, and nutrients. As discussed in Chapter 2, most soils are derived from rock and most nutrients are supplied; but with continuous cropping, nutrients are used and need to be replaced by plowing under green covered crops or applications of commercial fertilizer. Nitrogen is not derived from rock and usually needs to be replaced every year when heavily cropped. Main nutrient requirements are nitrogen, phosphorus, and potassium; and commercial fertilizers come ready to apply them in various ratios depending on crop needs.

The most favorable tillage for best crop growth is desired. This is determined by long-term crop yield data and water and soil conservation. Knowledge of pest control and recognizing the various insects, weeds, and plant diseases and knowing how to deal with these problems is paramount. If in doubt, get the benefit of experts at the Natural Resources Conservation Service (NRCS), USDA, extension service, or state university personnel for best tillage and soil conservation and fertilizer products. For pest and weed control practices, there are professional consultants, extension, university, USDA, and industry personnel that can help identify the pest or weed problem and recommend treatment. Basic knowledge of fertilizers and agriculture chemicals is desired. If irrigation is practiced on your ranch or farm, the NRCS can help set up the system and operation. Field leveling and proper water drainage and movement to fully irrigate the field and crop is a requirement. Preventing excessive runoff and movement of water to unwanted areas is to be avoided. Irrigation experts can help. A good working knowledge of the weather and its effect on crops should be studied. The rancher and farmer are highly limited by weather conditions and some procedures cannot be done or are delayed because of bad weather. Weather can make the difference between success or failure to plant, grow, or harvest a crop.

The length of the growing season and what crops, cultivars, or hybrids grow can be determined by local experts, farmers, and seed companies. Neighboring farmers will be more than happy to discuss the latest Roundup Ready crops or genetically-modified crop or conventional crop depending on your needs and taste.

I know this is not a requirement but being a good mechanic, knowledge of machinery, and the ability to repair equipment with welding skills are most helpful. As a boy growing up on a cattle and wheat farm in northern Idaho, I watched my father repair machinery and overhaul tractors and combine engines in the midst of spring work or harvest. Without these skills the crop could have easily been lost. Nowadays the equipment is probably more reliable and the manufacturers don't want you to tinker with it, but some basic knowledge of machinery is still very useful.

The modern farmer and rancher should have business skills and knowledge to care for his operation to make it profitable and cut costs where necessary. With this comes good recordkeeping so that mistakes or shortcomings this year can be corrected next year. The modern farmer should be aware of marketing procedures, local markets for his product, and some idea of the federal, state, and local policies of government and how they might affect his business. Knowledge of markets may help anticipate selection of future crops, cropping procedures, and what crop rotations may be profitable. Some study of the environment and wetlands that may affect farm operations may need attention.

The rancher and farmer needs to keep abreast of new developments in machinery, crop species, crop cultivars and varieties and hybrids, crop rotations, cropping systems, weed and pest control, new chemicals, new cultivation techniques, anticipation of future markets, and new technologies that may help his operation. He should be aware of his neighbor's successes and failures, being helpful to his neighbors and be a good neighbor. He should reward himself by doing some enjoyable things like woodworking, reading, hunting, fishing, or whatever is enjoyable.

How does one get ready to be a farmer or a rancher? The first requirement is a love of the land and nature and the desire to produce something useful like food products. The farmer desires to be his own boss, free and in nature. How does one get equipped to be a rancher or farmer? If you grow up on a ranch or farm and desire to stay, you are already equipped to handle it. It will take some effort to keep up with the latest technology and innovations. There are short courses offered at schools and universities specific for training. These courses can last a few days to months. There are also many one-day meetings to inform individuals of the latest developments and procedures in agriculture sponsored by extension, university and college or industry. There are also full degree programs in animal science, agronomy, entomology, agricultural education, plant pathology, agriculture business, and others with B.S. degrees available. Some farmers and ranchers have also finished M.S. and PhD programs useful to their profession and well-being in agriculture or related business.

Livestock and Animal Science

Much of the same knowledge to grow plants is needed in livestock management. Business knowledge, marketing, and government policies are still at the forefront as well as keeping good records of your operation. The livestock grower must also be aware of new developments and technology.

Knowing the best species or variety for your area or operation is paramount relative to markets and personal preferences. As stated earlier, being aware of animal waste and their proper disposal is required. Keeping animals out of streams and rivers and preventing animal waste in runoff water is desired. Knowing the life cycle, history, and conditions necessary for successful livestock and poultry operations is required for profit and a good reputation.

There are a number of sources for information to further the success of your operation. Visiting with feed store personnel, fellow growers, the county agent, livestock and poultry specialists from private practice, or the extension specialist can help. State college and university specialists or the USDA can also help. There are also individual meetings, short courses, college trainings, or degree plans at the university or college level that have great animal science and poultry programs.

Proper nutrition for livestock can be learned, keeping costs low. Having some veterinary knowledge or access to the vet is desired. Many farm operations grow crops and have livestock, and these operations can complement each other. For example, animal wastes can be used to fertilize crops; and crops can be grown to feed the livestock. Some of the land can be used to pasture poultry and livestock and provide roughage and good nutrition.

Individuals who love the land, raising livestock, and being self-employed might benefit from this lifestyle and enjoy some of the training and education programs available through reading, online, television, radio, land-grant colleges and universities, and industry programs.

Getting started or staying in farming can be a lifetime opportunity, requiring dedication and hard work; but like any profession it is not without risk.

Summary

The modern U.S. farmer is well-informed, usually successful, but operates on a sometimes limited profit margin. Successful farmers know their resources and limitations and make best use of these resources and knowledge to stay in business. The crop farmer must have knowledge of the local climate, soil and water resources, marketing and government policies, agricultural chemicals, and the best performing plants and/or animal species for his area. Having some knowledge of mechanics and/or veterinary science or access to these services is desired. Information is available for long-term or new farmers by reading, online, television, radio, local seed, feed and agricultural chemical stores, private consultants, the extension service, colleges and universities, and several USDA agencies. Short-term, medium- or long-term farmer training programs are available or even full college degree programs at many U.S. universities and colleges. A passionate love for farming and nature as well as a desire to raise crops and/or animals for fun and profit is a prerequisite.

CHAPTER 19

CONCLUSION AND THE FUTURE

The purpose of this book is to show where our food supply comes from and the difficulty and huge effort farmers, ranchers, educators, and researchers and many others spend in making it possible.

As pointed out by many people in agriculture and society, possessing a productive food supply allows others in society to pursue other needs, interests, and careers. Early humans spent most of their time acquiring food, shelter, and a safe place to reside. Great advances have been made in agriculture even in the last 50 years.

The task of agriculture and agricultural education and research is never complete because new pests or environmental conditions constantly emerge that require new or improved genetics in plants and animals, soil fertility and water use changes, improved nutrition needs, and many other unforeseen problems.

The problems in agriculture are not all negative because it provides many challenging jobs for farmers, researchers, foresters, horticulturalists, agronomists, wildlife specialists, food scientists, research scientists, industry personnel, educators, machinery and agricultural chemical personnel, agricultural engineers, fertilizer experts, and so on.

Work in agriculture is very interesting and requires expertise in many fields and subject areas. It is rewarding to develop a new disease-resistant crop, registration of a new environmentally-friendly pesticide, or an efficient machine to harvest apples or some other crop.

Agriculture is therefore the world's basic industry without which other industries would fail to survive or would be significantly hampered since agriculture provides food, clothing, shelter, and other basic human needs.

Our present success in U.S. agriculture is based on a long history of trials and errors, successes and failures. We have borrowed knowledge, plants and animals from our own and other areas of the world and have attempted to improve the plants, livestock, and wildlife acquired by selection, genetic manipulation, and/or improved cultural practices. Plants and animals are usually adapted to an area similar to where they have evolved; but by proper genetic changes and/or good cultural practices, improved productivity may be acquired in new or native locations.

Our land-grant and university systems in the U.S. of continuous education, extension, and research are the envy of the world. The need for new discoveries and technology in agriculture and the environmental sciences are acknowledged, provided through research, and are made known to society through education and the extension service. Agricultural knowledge and related areas are taught in schools and universities for students returning to the farm, for further training and education or for those in the business world. Industry also makes significant contributions to society and agriculture; and they work closely with government, the university system, ranchers, and farmers.

Agricultural research is especially important with adequate funding to make new discoveries and solve ongoing problems and to maintain or increase agricultural production and quality. By 2050 the world's population is projected to be 9 billion humans, a significant increase over about 7 billion in 2017.

What is required of agriculture to maintain our food supply and environmental integrity for the future?

First, the basic needs of humans on this earth are not going to change. The same nutritional requirements of humans and the plants and animals they depend on will change little. Attempts will be made to improve the nutrition, vigor, production, and attractiveness of the plants and animals we depend on through research and demonstration.

Human needs for carbohydrates, fats, protein, minerals and vitamins are obtained from plant and animal products they consume. Carbohydrates, starches, sugars, and fats all provide energy to humans; proteins aid in transport and storage of nutrients. Minerals for humans and animals include a long list found in Chapters 2, 8, and 12. Vitamins are either fat-soluble (A, D, E, K) or water-soluble (C & B-complex), and mention these nutrition requirements show our human dependence on the plant and animal kingdom for life.

In the process of supplying food, care must be taken to preserve the germplasm of plants and animals we depend on and the soil environment. Special attention to well-funded research includes all phases of agriculture, agricultural finance, and marketing. Food allergies and nutrition problems need further investigation. Seed banks established in different areas of the world to preserve crop, weed, and wild plant species are required (Chapter 1).

Training and educational opportunities must continue in all phases of agriculture to train farmers, ranchers, scientists, extension personnel, professors, and future leaders for all careers in agriculture.

Laws and government regulations should protect the environment, humans, and livestock but should not be so restrictive as to harm agricultural business or the family farm. Government involvement in fair market value and policies is necessary but may be too restrictive in what can be raised, planted, or grown. To a limited degree the marketplace should determine the crops and livestock produced.

The United States has abundant and excellent soil, favorable weather, farmers, policies and attitude for world agriculture. Hopefully free enterprise will be allowed to flourish even with many more people added to the population. The American farmer is highly capable of nearly feeding the entire world.

We have discussed some of the new technologies throughout this book that are being used. No-till operations, genetically-modified crops and animals, and computerized logging operations to name a few. The future will feature these and other new ideas making farming, ranching, and gardening more environmentally friendly, safe, secure, and productive. However, pests, soil, water, and environmental problems will always be present and a continuous challenge. However, with continuous monitoring, adequate funding and education, the U.S. will lead the world and flourish in agriculture productivity and quality.

Dr. Rodney W. Bovey was raised on a livestock and wheat farm in Northern Idaho. He received the BS degree in Vocational Agriculture and the MS degree in Agronomy with a minor in Agricultural Chemistry from the University of Idaho in 1956 and 1959 respectively. In 1959 he was hired by the University of Nebraska as Instructor in Agronomy and Project Leader of the Use of Aircraft in Agriculture. He also completed the PhD in Agronomy with a Botany minor at Nebraska in 1964. Dr. Bovey was employed by the U.S. Department of Agriculture-Agriculture Research Service from 1964 to 1994 as a Research Agronomist to develop weed and brush management on Southwest rangelands at Texas A&M University. He later served as an adjunct professor at Texas A&M, Department of Ecosystem Science and Management from 1994 to 2016. During his career, Dr. Bovey and co-workers published over 300 scientific papers and several books.

www.ingramcontent.com/pod-product-compliance
Lightning Source LLC
Chambersburg PA
CBHW051945150726
47999CB00004B/1256